Jona Maurer

Transactive Control of Coupled Electric Power and District Heating Networks

Karlsruher Beiträge zur
Regelungs- und Steuerungstechnik
Karlsruher Institut für Technologie

Band 21

Transactive Control of Coupled Electric Power and District Heating Networks

by
Jona Maurer

Scientific Publishing

Karlsruher Institut für Technologie
Institut für Regelungs- und Steuerungssysteme

Transactive Control of Coupled Electric Power
and District Heating Networks

Zur Erlangung des akademischen Grades eines Doktor-Ingenieurs
von der KIT-Fakultät für Elektrotechnik und Informationstechnik des
Karlsruher Instituts für Technologie (KIT) genehmigte Dissertation

von Jona Maurer, M.Sc.

Tag der mündlichen Prüfung: 28. September 2022
Hauptreferent: Prof. Dr.-Ing. Sören Hohmann
Korreferent: Prof. Dr. Kankar Bhattacharya

Impressum

KIT Scientific Publishing

Karlsruher Institut für Technologie (KIT)
KIT Scientific Publishing
Straße am Forum 2
D-76131 Karlsruhe

www.ksp.kit.edu

Print on Demand 2023 – Gedruckt auf FSC-zertifiziertem Papier

ISSN 2511-6312
ISBN 978-3-7315-1276-9
DOI 10.5445/KSP/1000155178

Abstract

The aim to decarbonize the energy supply represents a major technical and social challenge. The design of approaches for future energy network operation faces the technical challenge of needing to coordinate a vast number of new network participants spatially and temporally, in order to balance energy supply and demand, while achieving secure network operation. At the same time these approaches should ideally provide economic optimal solutions. In order to meet this challenge, the research field of *transactive control* emerged, which is based on an appropriate interaction of market and control mechanisms. These approaches have been extensively studied for electric power networks. In order to account for the strong differences between the operation of electric power networks and other energy networks, new approaches need to be developed. Therefore, within this work a new transactive control approach for Coupled Electric Power and District Heating Networks (CEPDHNs) is presented. As this is built upon a model-based control approach, a suitable model is designed first, which enables to operate coupled electric power and district heating networks as efficient as possible. Also, for the transactive control approach a new fitted procedure is developed to determine market clearing prices in the multi-energy system. Further, a distributed form of district heating network operation is designed in this context. The effectiveness of the presented approach is analyzed in multiple simulations, based on real world networks.

Kurzfassung

Die Dekarbonisierung der Energieversorgung stellt eine enorme technische und gesellschaftliche Herausforderung dar. Der Entwurf von Ansätzen zur Betriebsführung zukünftiger Energienetze sieht sich dabei zum einen der technischen Herausforderung gegenüber, eine enorme Anzahl von neuen Netzteilnehmern zeitlich und örtlich zu koordieren, um Erzeugung und Verbrauch in Einklang zu bringen und dabei gleichzeitig einen sicheren Netzbetrieb zu gewährleisten. Zum anderen sollten diese Ansätze idealerweise ökonomisch optimale Lösungen hervorbringen. Um dieser Herausforderung zu begegnen entstand das Forschungsfeld der *Transactive Control* Ansätze, welches auf einer geeigneten Verzahnung von Markt- und Regelmechanismen beruht. Diese Ansätze wurden bisher umfassend für rein elektrische Netze untersucht. Um den großen Unterschieden in der Betriebsführung zwischen Stromnetzen und anderen Energienetzen Rechnung zu tragen, bedarf es der Entwicklung neuer Verfahren. Daher wird im Rahmen dieser Arbeit ein neuer *Transactive Control* Ansatz für gekoppelte Strom- und Wärmenetze vorgestellt. Da dieser auf einem modellbasierten Regelungsverfahren beruht, wird zunächst ein geeignetes Modell entworfen, welches es ermöglicht, gekoppelte Strom- und Wärmenetze möglichst effizient zu betreiben. Zudem wird für den *Transactive Control* Ansatz ein neues Verfahren zur Bestimmung geeigneter Markträumungspreise, im sektorengekoppelten System, entwickelt. Außerdem wird in diesem Kontext ein neues Verfahren zur verteilten Betriebsführung von Wärmenetzen entworfen. Die Wirksamkeit der Verfahren wird an Simulationsergebnissen, basierend auf realen Netzdaten, analysiert.

Thanks

To Sören Hohmann for giving excellent classes, creating unknown levels of motivation during my time as a student at the KIT. Also, for providing the chance of pursuing this work and the great leadership all the way.

To Kankar Bhattacharya for taking the time to be my second examiner and for the great guidance, including long fruitful discussions during my research stay in Waterloo.

To Mathias, Claudio, Stefan, and Martin for great mentorship.

To Armin, Martin, Bertus, Pol, and Felicitas for reviewing this work.

To Jochen and Nicolai for sophisticated implementations.

To the Stadtwerke Kiel and Ettlingen, especially Silas Reigardt and Eberhard Oehler, for reviewing results and providing valuable practical insight.

To Kaori Nagato-Plum for her advice.

To Sophie for affectionate illustrations.

To my students for their motivation, questions, and ideas.

To the entire IRS staff, including Lennart Merkart, for thought-provoking impulses and companionship.

To my family and friends for ongoing support.

To Franzi and Kalea for being the best imaginable substitute colleagues during the pandemic creation times of this thesis.

Karlsruhe, August 2022

As the saying goes, the Stone Age did not end because we ran out of stones; we transitioned to better solutions. The same opportunity lies before us with energy efficiency and clean energy.

Steven Chu

Contents

List of Figures

List of Tables

Abbreviations and Symbols

Abbreviations

Abbreviation	Description
AC	Alternating Current
ADMM	Alternative Direction Method of Multipliers
AL	Augmented Lagrangian
APM	Alternative Power Method
ARH	After Rolling Horizon
BESS	Battery Energy Storage System
BRH	Before Rolling Horizon
CEPDHN	Coupled Electric Power and District Heating Network
CFVT	Constant Flow and Variable Temperature
CHP	Combined Heat and Power
COP	Coefficient of Performance
CPU	Central Processing Unit
DAE	Differential Algebraic Equation
DC	Direct Current
DER	Distributed Energy Resource
DHN	District Heating Network
DHS	District Heating System
DLC	Direct Load Control
DLMP	Distribution Locational Marginal Price
DMPC	Distributed Model Predictive Control
DPR	Differential Pressure Regulator
DSM	Demand-Side Management
DSO	Distribution System Operator
EB	Electric Boiler
EC	Energy Converter
ED	Economic Dispatch
EMS	Energy Management System
EMSSA	Energy Management System Software Agent
EPN	Electric Power Network
EPS	Electric Power System
EV	Electric Vehicle
FD	Finite Difference
FDP	Flow Direction Path
FNP	Flexible Network Participant

Abbreviation	Description
GFCP	General Form Consensus Problem
GMRES	Generalized Minimal Residual
HP	Heat Pump
HPA	Hybrid Pricing Approach
IEHS	Integrated Electricity and Heat System
IMO	Independent Market Operator
INP	Inflexible Network Participant
ISO	Independent System Operator
ISOEMS	Independent System Operator Energy Management System
LMP	Locational Marginal Price
LMPA	Locational Marginal Pricing Approach
MES	Multi-Energy System
MICP	Mixed Integer Conic Programming
MILP	Mixed Integer Linear Programming
MINLP	Mixed Integer Nonlinear Programming
MIQP	Mixed Integer Quadratic Programming
MISOCP	Mixed Integer Second-Order Cone Programming
MO	Merit Order
MP	Market Participant
MPC	Model Predictive Control
NLP	Nonlinear Programming
NM	Node Method
NMPC	Nonlinear Model Predictive Control
NP	Network Participant
OCD	Optimality Condition Decomposition
OFDP	Opposed to the Flow Direction Path
OPF	Optimal Power Flow
OTC	Over the Counter
PCGM	Preconditioned Conjugate Gradient Method
PV	Photovoltaic
QP	Quadratic Programming
RES	Renewable Energy Source
RMSE	Root-Mean-Square Error
SCED	Security Constrained Economic Dispatch
SOC	State of Charge
SOTA	State of the Art
TC	Transactive Control
TCS	Transactive Control System
TE	Transactive Energy
TESS	Thermal Energy Storage System
UC	Unit Commitment
UMP	Uniform Marginal Price
UMPA	Uniform Marginal Pricing Approach

Abbreviation	Description
UMPD	Uniform Marginal Price obtained from Dispatch
VFCT	Varying Flow and Constant Temperatures
VFVT	Variable Flow and Variable Temperature
VMFD	Varying Mass Flow Direction
WPP	Wind Power Plant
ZPA	Zonal Pricing Approach

Symbols

Symbol	Description
a	Coefficients used in Energy Converter (EC) equation or inequalities
$\boldsymbol{A}$	Node edge incidence matrix
A^{cross}	Cross section of a pipeline
$A^{\mathrm{dhn,sgn}}$	Flow direction dependent incoming/leaving node edge incidence matrix
$\boldsymbol{A}^{-}$	Node leaving edge incidence matrix
$\boldsymbol{A}^{+}$	Node incoming edge incidence matrix
A^{stor}	Surface of storage
$\tilde{\boldsymbol{A}}$	Real valued matrix
β	Component coefficient
β^{wf}	Positive welfare area
$\boldsymbol{B}$	Loop edge incidence matrix
B^{sh}	Shunt susceptance
$\tilde{\boldsymbol{B}}$	Real valued matrix
B^{bus}	Susceptance of nodal admittance matrix element
c	Bid/Offer price per unit
C'	Shunt capacitance per length and phase
$\tilde{\boldsymbol{c}}$	Real valued vector
c^{temp}	Temporal conversion factor
c^{ump}	Uniform Marginal Price (UMP)
c^{umpd}	Uniform Marginal Price obtained from Dispatch (UMPD)
c^{w}	Specific heat capacity of water
χ	Boolean approximator variable
d	Diameter
dx	Element length
δ	Sufficiently small parameter
$\boldsymbol{\Delta}^{\mathrm{cent}}$	Central search direction
$\boldsymbol{\Delta}^{\mathrm{dist}}$	Distributed search direction
$\boldsymbol{\Delta}^{\mathrm{dist,new}}$	Distributed search direction determined using a preconditioner
$\Delta\varepsilon$	Small positive parameter
Δh	Height difference

Symbol	Description
Δk	Time step intervall
$\boldsymbol{\Delta\lambda}^{\mathrm{cent}}$	Vector of change of dual variables within two iterations for the central case
$\boldsymbol{\Delta\lambda}^{\mathrm{dist}}$	Vector of change of dual variables within two iterations for the distributed case
Δp	Differential pressure
Δp_0	Differential pressure reference
Δp^{dev}	Differential pressure deviation
$\widehat{\Delta p}$	Predicted differential pressure
Δp^{pre}	Predifined differential pressure
Δp^{ramp}	Differential pressure ramping limit
Δp^{ref}	Reference value for differential pressure
$\Delta T^{\mathrm{max,dev}}$	Maximum temperature deviation
$\boldsymbol{\Delta x}^{\mathrm{cent}}$	Vector of change of optimization variables within two iterations for the central case
$\boldsymbol{\Delta x}^{\mathrm{dist}}$	Vector of change of optimization variables within two iterations for the distributed case
E	Heat quantity in storage
E^{meas}	Measured heat quantity in storage
ϵ	Tolerance value
ε	Pipeline position parameter
η	Efficiency
f	Objectiv function
f^{s}	Marginal demand bid cost function
$f^{\mathrm{d,dis}}$	Dispatched marginal demand bid cost function
f^{dhn}	Objective function for District Heating Network (DHN) optimization
$f^{\mathrm{dhn,rest}}$	Part of objective function for District Heating Network (DHN) optimization not resulting from Optimality Condition Decomposition (OCD) approach
f^{eco}	Economic objective
f^{δ}	Extension of objective function
f^{hydr}	Function defining nonlinear relation between mass flow and differential pressure over an edge
f^{rect}	Continuous differentiable approximation of rectangular function
f^{s}	Marginal supply offer cost function
$f^{\mathrm{s,dis}}$	Dispatched marginal supply offer cost function
f^{soft}	Soft constraints
$\boldsymbol{f}^{\mathrm{sys}}$	Nonlinear multidimensional system function
$\boldsymbol{g}$	Multidimensional function containing inequality constraints
g	Objective function
G^{bus}	Conductance of nodal admittance matrix element

Symbol	Description
G^{sh}	Shunt conductance
G'	Shunt conductance per length and phase
g^{acc}	Gravitational acceleration
γ	Pipeline position parameter
γ^{pc}	Propagation constant
$\boldsymbol{h}$	Multidimensional function containing equality constraints
$\boldsymbol{h}$	Vector of equality constraints
$h^{\text{obj,dhn,hydr,sn,bi}}$	Hydraulic border node equality constraint of other zone in supply network used within objective of current zone
$h^{\text{obj,dhn,therm,bi}}$	Thermal border node equality constraint of other zone used within objective of current zone
$\boldsymbol{I}$	Identity matrix
K	Valve flow factor
κ	Time steps since water mass entered pipeline
$\mathbf{KKT}$	Karush-Kuhn-Tucker (KKT) matrix
$\overline{\mathbf{KKT}}$	Approximated Karush-Kuhn-Tucker (KKT) matrix
$\overline{\mathbf{KKT}}^*$	Approximated Karush-Kuhn-Tucker (KKT) matrix at the optimal solution
$\mathbf{KKT}^*$	Karush-Kuhn-Tucker (KKT) matrix at the optimal solution
L	Length
$\mathcal{L}$	Lagrangian function
l	Storage losses
$L^{\text{ind}'}$	Inductance per length and phase
$\mathcal{L}^{\text{al}}$	Augemented Lagrangian (AL) function
$\boldsymbol{\lambda}$	Dual variable vector
λ^{cong}	Congestion component of Locational Marginal Price (LMP)
$\lambda^{\text{dhn,hydr,sn,bi}}$	Lagrange multiplier from hydraulic node equation of other zone border node in supply network
$\lambda^{\text{dhn,therm,bi,sn}}$	Lagrange multiplier from thermal node equation of other zone border node
λ^{energy}	Energy component of Locational Marginal Price (LMP)
λ^{loss}	Loss component of Locational Marginal Price (LMP)
$\boldsymbol{\lambda}^*$	Optimal dual variable vector
λ^{tot}	(Total) Locational Marginal Price (LMP)
m	Polynomial order
$\boldsymbol{M}^{\text{dhn}}$	Edge network participant incidence matrix of the district heating network
$\dot{m}$	Mass flow
$\widehat{\dot{m}}$	Predicted mass flow
$\dot{m}^{\text{pre}}$	Predefined mass flow
$\boldsymbol{M}^{\text{epn}}$	Bus network participant incidence matrix of the electric power network

Symbol	Description
μ	Component coefficient
$\nabla \boldsymbol{h}$	Jacobian matrix of the equality constraints
$\nabla \boldsymbol{\mathcal{L}}$	Jacobian matrix of the Lagrangian
n	Cardinality
n^{p}	Amount of time steps on the prediction horizon
ν	Kinematic viscosity
ω^{wf}	Negative welfare area
P	Real power infeed at bus i or by NP n
P^{flow}	Real power flow between two buses
$\widehat{P}$	Predicted real power
Φ	Heat power
$\widehat{\Phi}$	Predicted heat power
Φ	Preconditioner matrix
Φ^{ramp}	Heat power ramping limit
p^{pre}	Predifined pressure
P^{ramp}	Real power ramping limit
p	Pressure potential
p^{steam}	Steam pressure
ψ	Pressure loss coefficent resembling curvature of a pipeline
Q	Reactive power infeed at bus i or by NP n
$\widehat{Q}$	Predicted reactive power
q^{obj}	Objectiv function
Q^{ramp}	Reactive power ramping limit
$\boldsymbol{R}$	Control path edge incidence matrix
r	Pipeline wall roughness
Re	Reynolds number
$\mathrm{Re}^{\mathrm{crit}}$	Critical Reynolds number
ρ	Density
ρ_0	Reference density
ρ^{al}	Penalty Parameter of augemented Lagrangian
$\rho^{\mathrm{ocd},*}$	Least linear congervence rate of Optimality Condition Decomposition (OCD) approach
ρ^{w}	Density of water
$\mathbb{R}$	Real numbers
R^{stor}	Heat transfer coefficient of storage
R'	Resistance per length and phase
R^{therm}	Thermal resistance
R^{wm}	Auxiliary water mass parameter
$\mathrm{sgn}_{\Delta\varepsilon}$	Differentiable approximation of the signum function
σ^{fdp}	Flow direction on Flow Direction Path (FDP)
σ^{ofdp}	Flow direction Opposed to Flow Direction Path (OFDP)

Symbol	Description
$\mathbb{S}$	Set
S^{wm}	Auxiliary water mass parameter
T	Temperature
$t^{\max}$	Maximum length of stay of water mass in pipeline
t	Length of stay of water mass in pipeline
T^{a}	Ambient temperature
T^{bot}	Temperature at bottom of storage
t^{dead}	Dead time
T^{freeze}	Freezing temperature
$\widehat{T}$	Predicted temperature
T^{in}	Input temperature
T^{mean}	Mean temperature
T^{meas}	Measured temperature
T^{out}	Outlet temperature
T^{out1}	Lossless pipeline outlet temperature
T^{out2}	Lossy pipeline outlet temperature
T^{p}	Temperature of time step before simulation/prediction horizon
T^{ramp}	Temperature ramping limit
T^{top}	Temperature at top of storage
τ	Pipeline loss coefficient
$\boldsymbol{u}^*$	Vector of optimal control values
$\boldsymbol{u}$	Vector of control values
U^{ump}	Amount of power traded at Uniform Marginal Price (UMP)
U^{umpd}	Amount of power traded at Uniform Marginal Price obtained from Dispatch (UMPD)
V	Voltage amplitude
v	Flow velocity
W	Social welfare
w	Weight parameters in pipeline model
$w^{\mathrm{rect,h}}$	Weight influencing height of rectangular function approximation
$w^{\mathrm{rect,w}}$	Weight influencing width of rectangular function approximation
$\boldsymbol{x}_0$	Initial variable vector
x	Optimization variable
$\boldsymbol{x}$	Variable vector
$\boldsymbol{x}^{\mathrm{d}}$	Vector of power demand by all consumers
$\boldsymbol{x}^{\mathrm{s}}$	Vector of supplied power by all producers
$\mathcal{X}$	Real values multidimentional set
$\boldsymbol{x}^{\mathrm{st}}$	Vector of system state variables
$\boldsymbol{x}^{\mathrm{st,0}}$	Vector of initial system state
ξ	Friction factor
$\boldsymbol{y}^0$	Starting point for optimization
$\underline{\boldsymbol{Y}}^{\mathrm{bus}}$	Nodal admittance matrix

Symbol	Description
$\underline{\boldsymbol{Y}}^{\mathrm{f}}, \underline{\boldsymbol{Y}}^{\mathrm{ff}}, \underline{\boldsymbol{Y}}^{\mathrm{ft}}$	Auxiliary admittance matrix
$\underline{Y}^{\pi}$	π-equivalent admittance
$\underline{Y}^{\mathrm{sh}}$	Shunt admittance
$\underline{\boldsymbol{Y}}^{\mathrm{sh}}$	Shunt admittance vector
$\boldsymbol{y}^{*}$	Second-order Karush-Kuhn-Tucker (KKT) point
$\underline{Y}'$	Shunt admittance per length and phase
$\underline{\boldsymbol{Y}}^{\mathrm{t}}, \underline{\boldsymbol{Y}}^{\mathrm{tf}}, \underline{\boldsymbol{Y}}^{\mathrm{tt}}$	Auxiliary admittance matrix
$\underline{\boldsymbol{Y}}^{\mathrm{trl}}$	Transmission line admittance matrix
Z	Entering time step matrix element
$\boldsymbol{Z}$	Entering time step matrix
$\underline{Z}^{\mathrm{char}}$	Characteristic impedance
Z^{p}	Entering past time step matrix element
$\boldsymbol{Z}^{\mathrm{p}}$	Entering past time step matrix
$\underline{Z}^{\pi}$	π-equivalent impedance
$\mathcal{Z}$	Real values multidimentional set
$\boldsymbol{z}^{\mathrm{st}}$	Vector of algebraic state variables
$\underline{Z}'$	Serial impedance per length and phase
z	Optmization variable
$\boldsymbol{z}$	Vector of optimization variables

Operators

Operator	Description
$\widehat{\square}$	Estimated or predicted value
$\underline{\square}$	Complex value
$\{\cdot\}_i$	i-th Element of a vector
$\{\cdot\}_{i,j}$	(i,j)-th Element of a matrix
$\square^{-1}$	Inverse of a matrix
$\nabla\square$	Differential operator
$\overline{\square}$	Variable transformed into parameter
$\mathring{\square}$	Alias of index
$\rho(\square)$	Spectral radius
$\tilde{\square}$	Alias if applied to index
$\square^{\top}$	Transposed vector or matrix

Indices and Superscripts

Abbreviation	Description
0	Initial state, reference value
b	Boundary

Abbreviation	Description
cp	Control path
e	Edge
ev	Event
i, i'	Node/Bus
k, $\tilde{k}$	Time step
k^{p}, $\tilde{k}^{\mathrm{p}}$	Past time step
l	Loop
m	Polynomial order
n	Network participant
n^{dhn}	Network Participant (NP) of District Heating Network (DHN)
n^{epn}	Network Participant (NP) of Electric Power Network (EPN)
ν	Iteration of distributed optimization approach
σ, $\tilde{\sigma}$	Flow direction
σ^{-}	Flow against edge direction
σ^{+}	Flow in edge direction
ζ, $\mathring{\zeta}$	Position in pipeline, counted from the node where water masses enter the pipeline for a specific flow direction σ
$+$	Incoming edge, positive flow direction
$-$	Leaving edge, negative flow direction
a	Ambient temperature
al	Augmented Lagrangian
all	All
arh	After rolling horizon time step shift
be	Border edge
bi	Border node
bl	Border loop
bot	Bottom of storage
brh	Before rolling horizon time step shift
bus	Bus
ce	Cardinality of edges
cent	Central
char	Characteristic
chr	Charging
ci	Cardinality of buses
ck	Cardinality of time steps
crit	Critical
cross	Cross section
d	Demand
dchr	Discharging
δ	Extension of objective function
dev	Deviation

Abbreviation	Description
dhn	District Heating Network (DHN)
dist	Distributed
dpr	Differential Pressure Regulator (DPR)
epn	Electric Power Network (EPN)
exch	Heat exchanger
f	From
fdp	Flow Direction Path (FDP)
ff	From from
flow	(Line) Flow
fnp	Flexible Network Participant (FNP)
freeze	Freezing (temperature)
ft	From to
haa	Haaland
hydr	Hydraulic
in	Input of edge
inp	Inflexible Network Participant (INP)
max	Maximum
mean	Mean value
meas	Measurement
min	Minimum
new	New
np	Network Participant (NP)
obj	Objective
ocd	Optimality Condition Decomposition (OCD)
ofdp	Opposed to Flow Direction Path (OFDP)
out	Outlet of edge
out1	Lossless pipeline outlet temperature
out2	Lossy pipeline outlet temperature
p	Past
pc	Propagation constant
π	π-equivalent
pipe	Pipeline
pre	Predefined before simulation/optimization
pump	Pump
ramp	Ramping limit
ref	Reference
rest	Rest
rn	Return network
s	Supply
sh	Shunt
sn	Supply network
steam	Steam (pressure)
$*$	(At) Optimal point

Abbreviation	Description
stor	Storage
swamee	Swamee
t	To
tf	To from
therm	Thermal
top	Top of storage
trl	Transmission line
tt	To to
umpd	(Value at) Uniform Marginal Price obtained from Dispatch
vlv	Valve
vmfd	Variable Mass Flow Direction (VMFD)
w	Water
wm	Water mass
za	Zone a
zb	Zone b

1 Introduction

1.1 Motivation

To limit the impacts of climate change societies around the world are putting forward large efforts to decarbonize their energy supply. In order to maintain social acceptance political authorities thereby always seek to keep the financial burden as low as possible [GCM17]. Thus, the following four measures have been partly, and further need to be, implemented to achieve the aforementioned goals:
First, a **CO_2 free supply with electricity and heat** needs to be achieved. Therefore, Renewable Energy Sources (RESs) such as wind, solar, biogas, hydro, marine and geothermal power plants are integrated into energy networks in large scale to decarbonize the energy supply. Renewable energy accounted for 12.6 % of the total energy supply in 2020 and more than 10 % of the electric power demand in 2021 worldwide [REN22]. However, the energy production of RESs is strongly volatile [LLJZ20] and largely decentralizes the power infeed into the energy systems [ZBC04], which necessitates further measures stated below. The decarbonization of District Heating Networks (DHNs) is achieved by installing Combined Heat and Power (CHP) plants, Heat Pumps (HPs) and Electric Boilers (EBs) which are then run with green gas and electricity produced by RESs [NMZ+16]. This intensifies the interaction of Electric Power Networks (EPNs) and DHNs [CWW+18b]. This coupling is further increased by utilizing the waste heat originating from the cooling of supermarkets [GS21a], data centers [LSG+17], and industrial processes powered by electricity [PSH20]. The possibility of using this (decentral) waste heat is a major driver to further expand existing DHNs and design new forms of DHN operation [LSG+17]. Already nowadays, the utilization of DHNs reduces the CO_2 emissions in Europe by 113 million tons per year, which sums up to 2.6 % of the European CO_2 emissions [LSZW15]. Thus, DHNs will play a central role in future energy systems [MD17].

Second, the **provision of flexibility** through storages, power plants and Demand-Side Management (DSM) is mandatory to match energy supply and demand in the presence of strongly volatile energy supply of RESs. Therefore, storage systems[1] are more and more integrated in the EPNs. Nevertheless, storing electric energy with high efficiency, low costs, and ecologic compatibility is still a major issue [ARRB16], [GT17b]. However, storing thermal energy over long, even seasonal, periods at low costs is comparatively simple [XWL14] and thus thermal storages are an inherent part of facilities and DHNs [CMT+17, p. 76]. Thus, operators of DHNs have been incentivized to construct large water storages in the last years in Germany

[1] For example battery power plants, pumped storage hydro power stations or compressed air energy storages. Other concepts such as the usage of electric vehicles [DTKM18], gravity storages [Fyk19], underwater hydro storages [KAA+19], chemical storages [DDKB+20], and thermal storages [SS18] are also being tested.

[CMT+17, p. 79]. Further, a large potential is seen in utilizing the thermal water storages in facilities. For example in Germany 90 % of the present buildings using a heat pump are also equipped with a thermal water storage [ZWW+21]. These potentials need to be made utilizable by creating incentives for facilities by adequate jurisdictions and market frameworks [CN17]. General benefits arising from flexibility provision by DSM include less needed generator capacities, reduced generation costs due to less fossil fuels needed to run stand by capacities, and thus also less emissions [Str08].

Third, the **prevention of network congestions** is a major task in the EPNs as the existing networks were planed based on scenarios with different energy flows[2]. Therefore, grid reinforcements are necessary to adjust the EPNs to the current and future power flows. However, these take long planning periods and are very costly. Alone in Germany around 70 billion euros are necessary for the extension of the transmission network until 2035 [Bun19b, p. 4] and another 54 to 75 billion euros for the distribution networks until 2050 [Ago19, p.10]. Thus, flexibilities already named above should not only be used to match demand and supply, but also to prevent congestions and thereby decrease the amount of necessary grid reinforcements. In this context, several studies have shown the effectiveness of optimally coordinating demand and supply of grid participants to reduce network reinforcements [CN17, p. 30].

Fourth, **avoiding economic and technical inefficiencies** is of central importance. Three aspects will be highlighted in the following. The prior and purely technical aspects regarded, are the heat losses in DHNs. Heat losses mainly occur in DHNs, based on the heat flow through pipeline insulation. As heat losses increase with rising difference between the fluid and the ambient temperature, these can be minimized by keeping the temperatures as low as possible [ZLG13].[3] Therefore, DHN operators aim to set the mass flows through the heat exchangers of Flexible Network Participants (FNPs) such that the temperature difference between supply and return network is maximized, and thereby the mass flows and the resulting pump costs are minimized [ZLG13]. These mass flows are adjusted by pumps, Differential Pressure Regulators (DPRs), their respective control paths, and valves. As heat losses are in the range of 12 % to 20 % [VSLD13, ÇYÇ04, LMMD10], an efficient operation based on variable mass flows and variable temperatures is of significant interest.

A further aspect causing technical and economic inefficiencies arises when market and control mechanisms work separately or uncoordinated. Cases, where market mechanisms implemented without fully integrating the control mechanisms can be found in electricity markets

[2] Thereby, it was assumed, that energy is produced in a few large central plants and then distributed to the consumers over hierarchical structured networks. However, in the meantime the majority of newly connected RESs feed power into distribution networks instead of transmission networks [CN17, p. 29]. Additionally, novel consumers such as HPs and Electric Vehicles (EVs) consume more electric power and do so simultaneously in the same regions. Also, different RESs types, as solar panels and Wind Power Plants (WPPs), each generate regional correlations in their respective power supply due to similar meteorological conditions. The interplay of these effects as well as new regulations enabling competitive energy markets with numerous new decentralized market participants creates a rising amount of network congestions in EPNs [PKK15].

[3] Supply network temperatures can only be reduced to a certain level in order to sufficiently supply the heat demand of customers.

cleared by uniform pricing. For example, in the European energy sector, electric energy markets are operated based on uniform marginal pricing[4]. Whenever the power flows resulting from their allocation lead to network congestions in Germany, a highly regulated redispatch procedure takes place which leads to uplift costs. Grid operators order redispatch and select suitable power plants. The units that are shut down receive compensation, while those that are started receive a corresponding payment. These additional costs are passed on to all electricity consumers as network fees [Bun18b]. The costs for these redispatch measures[5] have risen in the past decades and exceeded 1.4 billion euros in 2020 in Germany [Bun22]. Also, the high amount of necessary command and control instructions has led to very complicated scenarios for system operators, in networks operated by these markets mechanisms [MSC03].

Last but not least, the aspect of techno-economic issues created by a separate operation of each domain of a Multi-Energy System (MES)[6] has to be kept in mind. Energy converters that are connected to DHNs and EPNs as e.g. PtH plants, or Combined Heat and Power (CHP) plants, are usually operated heat-driven or power-driven[7]. This form of operation and the concomitant form of pricing of energy, leads to cross-subsidies from one energy network to the other. A well known example of this phenomenon is the alternative power method, where the prices in DHNs are dependent on the outcome of electricity markets [DLS+19]. More general, when heat markets and electricity market clearing processes affect each other, for example due to energy converters which take part in both markets, simultaneous clearing with instantaneous communication between both markets is of interest to obtain an overall optimal result. Otherwise, in the case of sequential clearing, the first clearing process is dependent on predicted information of the second clearing process, and the feasible region of the second market clearing could be unfavorably constrained [ZWW+21]. Further, a pure heat-driven or power-driven operation of energy converters neglects the ability to counteract contingencies in one of the connected network types. Besides, flexibilities existing in the DHNs, as large water storages [ZWW+21], pipeline storage capabilities [Ick95] and DSM can not be optimally utilized to support the EPN in matching supply and demand. Also, due to a lack of coordinated operation of DHNs and EPNs, wind power is curtailed in many cases which could have been used in PtH devices, to supply heat to DHNs [DF14, MKP20]. In addition to the aforementioned, Multi-Energy System (MES) also exploit better conversion efficiencies and increase the environmental performance [Man14]. In summary, the benefits resulting from a joint operation of Coupled Electric Power and District Heating Networks (CEPDHNs) and their markets are well documented [ZWW+21, WYJ+16, LSF14, DLS+19].

---

4 Within every price zone.

5 Including feed-in management, countertrading as well as procurement and usage of network reserves.

6 Besides MES, further terms used in the same context are integrated energy systems [WYJ+16], multi-carrier energy systems [CWW+18a], and sometimes energy internet [ZLGZ17].

7 These two forms of operation are also known as following thermal load or following electric load in the literature, see e.g. [LWW18].

Based on the above stated challenges in CEPDHNs, the following **beneficial properties** can be deduced to achieve a technically efficient and economically optimal operation of future CEPDHNs:

1. The ability to coordinate the vast rising number of flexible network participants in EPNs and DHNs spatially and temporally in real-time. Thereby, matching supply and demand while ensuring secure[8] CEPDHN operation and keeping the necessary grid reinforcements as small as possible.

2. Technical efficiency is to be increased. Especially, a form of DHN operation with low losses is to be named in this context.

3. Market and control mechanisms are to be designed jointly in order to interact optimally. For that matter, inefficiencies as uplift costs stemming from redispatch procedures can be prevented, by including operational network constraints within the market clearing processes.

4. EPNs and DHNs and their energy markets are to be operated in a coordinated form, in order to make use of the full gain of technical, economic and environmental performance.

1.2 State of the Art

To face the interdisciplinary context of the presented work, four different fields of research spanning optimal operation of CEPDHNs are stated here. It should be noted, that these research fields partly overlap, since they all deal with the same aim of optimizing FNPs in CEPDHNs. However, the individual approaches differ largely, due to assumptions on information sharing between entities as network operators, market operators, and FNPs, the regarded form of network operation, as well as the chosen objectives as e.g. cost minimization or social welfare maximization. The first group of papers tackles the cost minimizing operation of CEPDHNs. These approaches model cost minimizing scheduling and dispatch as performed by system operators. The second group of papers focuses on distributed operation approaches for CEPDHNs with the aim of enabling real-time implementation of large optimization problems through parallelization and/or protection of confidential information of system operators running connected and interacting systems. The third field of interest considers approaches for market-based operation of CEPDHNs. These papers focus on the market aspects and mechanisms relevant in the context of energy markets. The fourth and last cluster of the state of the art describes the most relevant and recent work within the field of Transactive Control (TC). The domain free concept of TC brings together market and control mechanisms to attain global objectives within energy system operations. Finally, the last part of this section gives a discussion of the content of the presented literature and describes the existing research gap. This comprises the missing full integration of market and control mechanisms, the insufficient CEPDHN modeling detail for efficient operation, and the lack of scalability of most optimal CEPDHN operation approaches.

8 All network states are within operational boundaries provided in Section 3.3.1.

1.2.1 Cost Minimizing Operation of CEPDHNs

The approaches presented within the section have been developed for cost minimizing scheduling and dispatch as performed by system operators. In this context, market mechanisms as e.g. different forms of pricing which are necessary in deregulated markets, are neglected. The operations mostly differ by the used CEPDHNs model, which lead to diverse classes of optimization problems solved by different methods. The following works are characterized by a high technical detail in comparison with the ones outlaid in the other three following subsections.

The authors of [GA07] identified the need for MES modeling and operation early and presented relevant work in the field of optimal dispatch of MES. Therein, they used purely stationary network models and the energy hub concept. This concept is also applied by the authors of [BHH$^+$12] to provide DSM[9]. Herein mathematical models for different household devices and their respective demands are presented. These are used to formulate the optimization problem of a residential micro hub, with which different objectives as e.g. cost minimization or emission reduction can be pursued. The overall optimization problem of the residential micro hub is thereby formulated as a Mixed Integer Linear Programming (MILP) problem. The approach is demonstrated in simulations and with real prototypes. In this work, the heat loads are solely supplied by household devices. Energy hubs are a possible form of structuring models of entities as e.g. distribution networks, facilities or dwellings. Typically the power flows within the energy hubs are modeled as lossless and the power conversion between different energy carriers are modeled by conversion efficiencies. As this concept neglects operational constraints of the multi-carrier networks within the hubs and creates modeling overhead in the context of this work, it is not further regarded here.

An approach for the heat and power dispatch in a CEPDHN is proposed in [LWS$^+$15]. Thereby, the operation costs of CHP units, of thermal plants, and the curtailment of wind power is minimized. The EPN is modeled based on a Direct Current (DC) power flow model, neglecting the reactive power flows and voltage magnitudes of real Alternating Current (AC) operation. The pressure losses within the DHN are calculated based on constant friction factor coefficients within the Darcy–Weisbach equation. Also, loops within the DHN are neglected, as no constraint is used to fix the pressure drops within a loop of the network. Mass flows throughout the network are modeled as being variable, whereas the mass flow direction remains fixed. To represent the thermal dynamics of the temperature fronts within the DHN, the authors base the thermal pipeline model on the node method [Ben91], which is presented in detail in Section 2.3.3. This results in a Mixed Integer Nonlinear Programming (MINLP) problem. To circumvent the MINLP, the authors propose a decoupled approach, where the complicating variables[10] are determined based on the results of the previous optimization. The solution is then obtained by several iterative optimizations and recalculations of the complicating variables. The authors assure, that this procedure shows good convergence in practical applications, even though a proof of convergence is still an open field of research. Information on

[9] An overview on the different forms of DSM, as energy efficiency, time of use, demand response, and spinning reserve is found in [PD11].

[10] These are integer variables here.

necessary Central Processing Unit (CPU) time for the calculations is not given. However, this work shows, how the storage capacity of the DHN pipelines can be used effectively to increase the economic efficiency of the entire CEPDHN operation.

The MINLP problem from [LWS+15] is reduced to a MILP problem in [LWW+15] by assuming a constant flow and variable temperature control (CF-VT) operation form. This approach is used to obtain the cost minimizing Unit Commitment (UC) solution of the regarded CEPDHN for different problems. These are a pure deterministic UC calculation of the entire CEPDHN, a separate master slave optimization of the EPN and the subsequent DHN optimization based on benders decomposition, and a robust optimization problem, which enables to take uncertainties of the wind power infeed into account.

A further unit commitment approach for optimal operation of a CEPDHN is presented in [GWL+17]. In this work, the thermal inertia of buildings is also used to lower the operation costs of the CEPDHN and to achieve a higher utilization rate of the wind power. Therefore, dynamic models are introduced to considered the thermal storage and loss of the buildings. The proposed MINLP problem approach is also based on the Constant Flow and Variable Temperature (CFVT) control mode assumption for the DHN and the EPN is modeled solely by a real power balance equation, omitting any further constraints as e.g. flow constraints. A case study of a real world radial CEPDHN is performed to show the functionality of the approach. CPU calculation times are not given.

In [WGJ+17] the fuel cost minimizing scheduling of a CEPDHN is performed by solving a MINLP. For that matter, the optimal interaction of CHPs, EBs, Thermal Energy Storage Systems (TESSs), and consumers is pursued. The authors describe the presented approach as a form of Model Predictive Control (MPC). The DHN is operated by the CFVT control strategy. Hence, all mass flows, incorporating their flow directions, are defined in advance while the temperatures are dependent on the MPC control signals. The EPN is modeled by linear power balance and flow constraints.

An integrated heat and electricity dispatch model is developed in [ZZZW18b], and implemented to minimize the daily operation cost as well as the wind power curtailment in the CEPDHN. The DHN pipelines are modeled based on the node method. The DHN is operated based on a special CFVT form, where the flow can be varied in advance for different time steps. The EPN is represented by a linear real power balance equation. Based on the aforementioned simplifications, the overall dispatch optimization problem is reduced to a large scale linear programming problem. This approach was extended in [ZZZW18a], by taking into account variable mass flows and the possibility of utilizing the thermal building inertia. As the auxiliary variables necessary for the node method seem to be predetermined in some way, the problem still remains a large scale linear programming problem. No information on the accuracy of the proposed models is given. Also the flow directions in the pipelines of the DHN are fixed, which is surprising, due to the variable flow control and the amount of producing units in the case study. However, the paper gives a good overview on the basic forms of DHN operation.

The nonlinear unit commitment problem solved in [LWW+19], with nonlinear constraints, is approximated to obtain a Mixed Integer Quadratic Programming (MIQP) problem. This is then solvable in a few seconds for a simple test system and within 15 minutes for larger test systems, considering uncertainty. However, no information of the model accuracy, e.g. by giving a comparison of the model output with measurement data is given. Further work with similarities to the above mentioned is found in [ZLZ+19, YCC+19].

A good overview on the modeling and solution methods for operational optimization of CEPDHNs, thereby named the Integrated Electricity and Heat System (IEHS) is given in [ZWW+21]. Important conclusions, which are given in this paper and concur with the analysis of the author of this work are: First, "the electrical, thermal, and hydraulic processes are tightly coupled in the IEHS, leading to a large computational burden for the integrated energy flow calculation"; second "nonlinear terms or non convex feasible region caused by device characteristics, network constraints, and optimization under uncertainty make the original problem computationally intractable"; third, "the time delays and thermal dynamic characteristics in the pipelines of the DHS are non-negligible"[11].

The central assumption which strongly facilitates the original problem of efficient DHN, and thereby also CEPDHN, operation comprises a constant mass flow [HTWL19]. By consequence, this brings along a second assumption which is typically not valid in efficient DHN operation: Mass flow directions are predetermined before solving the optimal operation unit commitment or dispatch problem. This fundamental aspect is explained in the following Remark.

Remark 1.1:
As soon as more than one point of heat power infeed exists or the DHN has a meshed structure within the supply or return network[12] ***Varying Mass Flow Directions (VMFDs)*** *can occur during network operation [Opp15, p. 75]. This is due to the fact, that the flow directions on DHN edges is dependent on valve and pump operation set points in DHNs, and these set points are varied by the network operators, producers and consumers.*

To the authors knowledge, the only two approaches published, which do not use these assumptions for optimal CEPDHN operation are found in [Trö99, HTWL19]. In [Trö99] an approach for the optimal dispatch of the producers of a DHN operator is presented. The chosen objective is to minimize the necessary energy purchase. Thereby, the interaction with the EPN is regarded to model the amount of produced electricity to enable economic assessment of different solutions. The approach enables to incorporate the storage capabilities of the DHN pipelines while regarding multiple producers at different locations within a meshed network. To represent the temperature dynamics in the case of VMFDs a pipeline model based on the linear interpolation of node temperatures as in [Ben91] is presented. However, this approach assumes a uniform flow direction for the entire prediction horizon, which creates deviations of the pipeline outlet temperature, when this assumption is not fulfilled. Necessary integer variables are approximated by continuous variables, utilizing additional nonlinear constraints. Also not continuous differentiable constraints, as e.g. with absolute value operators, are approximated by nonlinear continuous differentiable approximations. To reduce the computational burden of the proposed approach, stationary DHN models are used to calculate initial solutions. The non stationary thermal variables of the stationary solution are then recalculated based on a dynamic simulation model. The resulting values are then utilized as the starting points for the overall Nonlinear Programming (NLP) problem.

[11] District Heating System (DHS).

[12] DHNs are typically constructed symmetrically. However, it is important to note that the meshes occur within the supply or return network as, when taking both network parts into account, every DHN is of meshed structure.

The approach in [HTWL19] is proposed for economic dispatch of a CEPDHN , where VMFDs are taken into account in the DHN . Therefore, first a complex non-convex MINLP model is established, which is then approximated by a Mixed Integer Conic Programming (MICP) problem. This new convex model problem can then be solved efficiently by standard of the shelf solvers. The initial nonlinear model is of stationary form and thus disregards the propagation of the temperature fronts through the DHN pipelines. Also the friction factor in the Darcy-Weisbach equation is modeled by a constant coefficient, which is a strong assumption as will be shown in detail in Section 2.3.2. Further, if storages are charged or discharged has to be defined before the optimization, as this defines their outlet temperature. The EPN is modeled by a DC line flow model. Determining the accuracy of the proposed approximated model is left for future work.

1.2.2 Distributed Operation of CEPDHNs

The literature summarized within this section is based on multiple distributed controllers operating CEPDHNs. Note, that in this work the terms **distributed** and **decentralized**, are defined as follows[13].

> **Definition 1.1**
> *A distributed control approach contains multiple local controllers which share information, while no centralized authority is in charge for this task. In contrast to that, decentralized control comprises subsystems, which are controlled independently by not communicating controls [Lun14a].*

The motivation for distributed operation approaches typically comes from at least one of the following five reasons: First, the large number of FNPs, needs parallel computation in order to solve the resulting large scale optimization problems. Second, the burden of communication increases with larger distance and amount of the individual entities. Third, a central control unit would need frequent redesign, as a change in a single element would require adaption. Fourth, information exchange may be nontrivial due to unwillingness[14] or lack of means, also aspects of cybersecurity become more relevant if data is accessible at many different points. Fifth, a central controller is more vulnerable, due to a single point of failure [YMS14, MDS^{+}17].

Early work in this field was published in [GHVO96]. Therein, the heat and power dispatch are performed in two sub problems, which are coupled through the feasible region constraints of possible heat-power production of the co-generation units. Through the exchange of information the two sub problems converge to the global solution. As well the EPN and the DHN are represented by power balance equations.

[13] Throughout the literature concerning the optimal operation of CEPDHN, the terms **distributed** and **decentralized** are used by different authors describing the same ideas. This results from the fact, that different criteria, as e.g. demand for communication or the location of necessary hardware, is utilized to decide if an approach is decentralized or distributed. Within this work we use the definitions coming from the control theory background [Lun14a]. However, this definition is also used frequently in power engineering, see e.g. [YMS14].

[14] For example, the EPN and the DHN can belong to different operators [ZWW^{+}21].

In [Arn11] a Distributed Model Predictive Control (DMPC) approach is presented to solve the optimal dispatch of energy hubs in different control areas. The power and gas networks between the hubs are represented by stationary models as in [GA07]. Heat flows and the system dynamics representing storage devices are considered only within the hubs. The Optimality Condition Decomposition (OCD) approach is used to decompose the global dispatch problem into a distributed multi-area optimization problem.

In [LFZC16] a new nonlinear CEPDHN dispatch model is presented together with a new distributed algorithm for the calculation of the optimal solution of the distributed dispatch of the DHN and the EPN. The EPN model is based on the accurate AC power flow model. However, the DHN model uses several new component models, while a validation of the model is missing. The general convergence properties of the developed distribution approach are not given.

A combined heat and power ED approach is given in [LWZS17]. Based on a constant mass flow and constant load assumptions for the DHN and a DC power flow model the overall ED problem can be formulated as a Quadratic Programming (QP) problem. The central problem is solved in distributed form by the DHN and EPN operators using benders decomposition. In every iteration, the EPN operator will calculate boundary variables, by performing an optimization, and communicates these to the DHN operator. Then, the last mentioned calculates the optimal DHN dispatch, based on constant boundary variables. These results are then used, in different forms in case of feasibility or infeasibility of the DHN dispatch, to initialize the next iteration of the EPN dispatch. Based on this, the solutions converge to the same solution as the central problem. The authors show, that the benders decomposition performs better than the Alternative Direction Method of Multipliers (ADMM) in their case study. However, the central problem calculation converges faster in the large scale problem in the presented simulations. This could arise from the fact, that benders cuts can offer few information on the central problem, which can result in weak convergence properties [LGZ+19].

The contributions of [HLW17] are threefold. First, the coordination framework for a CEPDHN is proposed with multiple DHNs and a single EPN, with their respective operators. Second, a convex economic dispatch optimization problem for the overall CEPDHN is formulated. Third, a distributed solution scheme, based on Optimality Condition Decomposition (OCD) is introduced. For that matter, the EPN and DHN operators communicate real power flow values over the tie lines, the phase angles of their respective buses and the Lagrange multipliers of the complicating constraints. The decomposition over electric tie lines has already been performed before, e,g. in [Arn11][15]. As the central ED optimization problem is convex, convergence of the OCD procedure is guaranteed. However, the convexity is only achieved, by assuming constant mass flows and a DC power flow model.

The distributed optimization in the following papers [XWZ+19, LGZ+19, TWW+20] is based on the Alternative Direction Method of Multipliers (ADMM). All of these papers use the constant mass flow operation form for the DHN operation. The EPN models are based on linearized AC branch flow models [XWZ+19], real power balance equations [LGZ+19], and a DC power flow model [TWW+20]. The entire non-convex MINLP CEPDHN scheduling

[15] Herein, different variables are shared among the areas, due to a slight different formulation of the problem

problem proposed in [XWZ+19] is reformulated as a Mixed Integer Second-Order Cone Programming (MISOCP) by replacing the nonlinearities by approximations. This problem is then decomposed into sub problems using the ADMM approach. To avoid the non convexities coming from the remaining integer variables, the MISOCP sub problems are then solved by further decompositions. The authors of [TWW+20] performed multiple optimizations with different time step intervals in order to understand the computational impact of this parameter choice. Based on their calculations, they propose 30 to 20 min as a good time interval for their multi period optimal CEPDHN dispatch, solved with the ADMM parallel computation approach. In [TWW+20] a Locally Adaptiv ADMM (LA-ADMM) approach is introduced to improve the convergence properties of the basic ADMM algorithm. No information is given on the accuracy of the proposed models used in the three previously discussed papers [XWZ+19, LGZ+19, TWW+20].

1.2.3 Market-based operation of CEPDHNs

The presented works in the following focus on the economic issues arising in optimal coordination for CEPDHN operation. Thereby, game theoretic methods, emerging market equilibria, arrangement of pricing mechanisms, and the interplay of DHN and EPN market operators are designed and analyzed.

Aggregators are envisioned in [SDWS17] to supply consumer heat and power demand at low prices. Therefore the aggregators are enabled to adjust the consumers demand[16] in order to minimize the energy purchase costs. Further it is assumed, that the entire CEPDHN is operated by a single Independent System Operator (ISO). The ISO maximizes the social welfare based consumer and aggregator bids. The entire energy dispatch, modeled as two-level optimization problem, is analyzed in case studies showing the efficient integration of wind power through flexible demand. The authors assume constant supply temperatures and variable mass flows, using a stationary DHN pipeline model and the EPN is modeled as a single node by a power balance equation.

The energy trading market equilibrium in a CEPDHN is determined in [CWL+18]. Optimization problems are developed for the analysis of the market impact of strategical behavior of producers and flexible consumers. The power distribution network market is cleared based on an AC flow model, which takes losses into account. For the DHN market clearing a CFVT DHN operation mode is used.

Similarly, in [LWM+18] a mathematical program with equilibrium constraints is introduced to analyze the strategic behavior of energy hubs in deregulated day-ahead heat and power markets with game theoretic methodology. In this context, the authors propose to use the pay-as-bid pricing method. The DHN model is based on a linear thermal flow model resulting from the constant mass flow assumption. The EPN representation is based on a linearized branch flow model.

[16] Also known as Direct Load Control (DLC) [PD11].

The parallel distributed dispatch of a DHN and an EPN, which are coupled to a CEPDHN is presented in [CWW+18b]. The optimal thermal flow problem of the DHN incorporates variable temperatures and variable flows. Also VMFDs can be represented in the model, but not within the DHN dispatch. The market driven approach presented, determines the prices, here Locational Marginal Prices (LMPs), by which the EPN and DHN operators sell heat or power to the other operator. Also the electric and heat power production of CHPs and HPs, owned and operated by either of the two operators, is determined in this fashion. Due to convergence problems, the authors utilize a continuous LMP demand curve approximation to overcome occurring oscillations. The approach exchanges similar information between the two network operators as the OCD approach in [HLW17], but the complicating constraints are not relaxed to the objective functions of the other operator[17]. Thus, this additional information, used to calculate the search direction, which is used in the OCD approaches, as depicted in detail in Section 3.3.2, is left out here.

In [DLS+19] the Locational Marginal Pricing Approach (LMPA), which has been extensively studied and used for electric energy markets is extended to heat markets. More detailed, the effect of transmission delay of DHN pipelines on the LMPs is analyzed. The authors conclude, that the LMPA can be also used for DHN pricing, advancing its well known benefits to the entire CEPDHN. The authors use a constant flow assumption for the DHN within their work.

The work in [MKP20] and [ZH21] propose a sequential clearing of the heat and power markets. In [MKP20] the heat market optimization problem takes the electricity market clearing into account, in order to obtain a high level of coordination in the clearing of the two independent operated markets. The authors developed this approach as a trade-off between state of the art uncoordinated market clearing and a fully integrated heat and power market, which would need very strong adaption of jurisdictions based on the authors estimate. The decomposed clearing is achieved by utilizing the benders decomposition. Based on case studies, the authors show, that the stronger market coordination enables a maximization of RESs power infeed. In contrast to this, the authors of [ZH21] propose to clear the power market before the DHN operators decide on their optimal schedule. Therein, price incentives are used to motivate the DHN operators to fully exploit the pipeline storage capabilities. The optimal thermal flow problem used for the DHN optimization is based on the CFVT operation mode.

The approaches presented in the first three fields of literature above, bring along a stronger focus on either the technical operation or the market mechanism design. An approach that regards both aspects in an integrated form is TC.

1.2.4 Transactive Control

This section, presents the most relevant literature in the field of TC. In the context of the aforementioned challenges, see Section 1.1, neither control nor market mechanisms alone can provide the essential basis for efficient economic and secure[8] CEPDHN operation. In this context, the concepts of Transactive Energy (TE) and TC have been developed [KW16]. These are defined as:

[17] Even though the Lagrange multipliers, here named LMPs, are used within the objective functions of the opposed network operator.

Definition 1.2

Based on [Cou15], TE is "a system of economic and control mechanisms that allows the dynamic balance of supply and demand across the entire electrical infrastructure using value as a key operational parameter."

Definition 1.3

The concept of TC [LLJZ20] is "a domain-free approach that integrates market-based coordination and value-based control for a group of resources to achieve certain global objectives."

Thereby, value is usually understood as market price [LLJZ20]. In this sense, TC enables, optimal coordination of FNPs while considering privacy[18][KW16, LLJZ20], overcoming the disadvantages of earlier concepts such as direct load control and price-responsive control, as the former does not consider user preferences and the latter lacks predictability of the load response [LLJZ20]. When TC is applied to an EPN, it is usually referred to as a TE system [LLJZ20], studied in a large number of papers, as summarized in [LLJZ20, ALFV19, LWHZ17]. As stated in [LLJZ20] TE, and therefore also TC, is characterized by: "the following distinctive features":

1. The class of approaches achieves end users participation and engagement "through smart energy management systems".
2. TE enables scalability of the system.
3. The framework provides "control signals that coordinate self-interested users to achieve system-level objectives while respecting power system constraints".
4. Within TE market interactions coordinate supply and demand sides.

These features distinguish TC approaches from the three aforementioned fields of research for optimal CEPDHN operation (cost minimizing operation, distributed operation, and market-based operation). Still, some of the distributed operation approaches presented in Section 1.2.2 try to provide scalability and most of the market-based operation approaches discussed in Section 1.2.3 take bidding consumers into account. But none of the presented approaches of the aforementioned fields brings together all of these distinctive features.

Due to the fundamental difference of the underlying physics, in network operation of EPNs and DHNs, such as relatively slow propagation of energy and the potential to store energy within DHNs, it is not possible to apply TC approaches developed for EPN operation to DHNs. Furthermore, as opposed to TC for EPNs, TC approaches for CEPDHNs have not been fully studied, as the application of TC techniques to multi-energy systems is relatively new. For example, in [BCDC18], thermostatic loads of heat pumps and air conditioners are operated based on the TC paradigm, while maintaining comfortable temperatures. A TC modeling and

[18] The aspect of privacy regarded here is [KW16]: "the approach protects the end user's privacy as the bids communicate only information about energy quantities and prices."

assessment framework for distributed multi-energy systems is presented in [GCHM19]. The authors of [WJH+19] propose a double auction retail market framework to enable optimal supply and consumption with electricity and heat. A form of the ADMM is used in a TC system to coordinate the distributed energy sharing among multi-energy micro grids in [YHA+20] and [XZL+20]. A double stage stochastic approach, which allows thermal energy storages of buildings to manage the uncertainty resulting from energy procurement in a TC context, is reported in [YP21]. Finally, a TC approach based on peer-to-peer transactions in the context of multiple energy hubs, with multiple energy carriers and high infeed of RESs, is discussed in [GYYW21].

However, these papers do not discuss issues associated with market-based operation that considers the relevant physical aspects of efficient CEPDHNs operation. In general the presented TC papers use very simple CEPDHN models, mostly neglecting thermal flows within the DHNs.

1.2.5 Discussion and Research Gap

A discussion of the above presented literature and a derivation of the research gap thereof is provided within this section.

Existing models for optimal operation of CEPDHNs suffer from unjustified simplifications. The following assumptions are questionable in real world examples:

- The operation form of the DHN guarantees constant mass flows. Thus these are known before the optimization and the hydraulic models do not need to be incorporated into the optimization problem as hydraulic and thermal models are independent [ZWW+21].
- Friction factors in the Darcy Weisbach equation, describing the pressure loss within DHN pipelines are modeled as parameters. This assumption is valid if the mass flows are constant [LWZS17] and thus predetermined. However, this model is also widely found when flows vary during the operation. In the latter case, the pressure difference calculation of the components becomes very inaccurate, see Section 2.3.2.
- Varying Flow and Constant Temperatures (VFCT) DHN operation mode is used. Therefore, potential heat loss reductions and temperature storage in DHN pipelines are neglected.
- The mass flow directions within the DHN are fixed and known in advance. However, this assumption is invalid as soon as more than one point of heat power infeed exists or the DHN has a meshed structure within the supply or return network[12], also see Remark 1.1. Thus this assumption is invalid for most real world DHNs.
- DPRs, pumps and their respective control paths are not incorporated into the hydraulic DHN models. Hence, the pressure differences over the components and the mass flows become inaccurate, which then also impact the thermal variables.
- Stationary thermal pipeline models are used in DHN models.

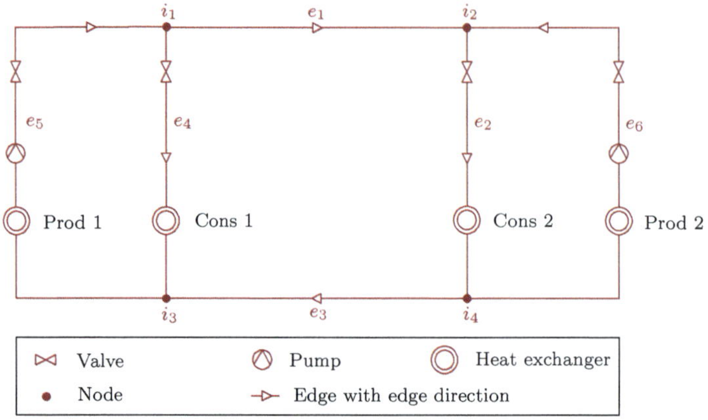

Figure 1.1: Example DHN with four nodes, two consumers and two producers.

- A radial and lossless DC EPN model is used. These assumptions help to avoid nonlinear and non-convex AC power flow equations. Still, electric distribution networks line losses should be taken into account [CWL+18] and meshed structured distribution networks exist as well.

Using the aforementioned simplifications enables to avoid non-convex and nonlinear constraints and integer variables within the optimization problems. However, not considering relevant physical effects may have severe consequences on technical efficiency and the market outcome. Exemplary direct consequences of the assumptions above on technical efficiency and market clearing are:

- Constant mass flow: In this case the valves and pumps are all modeled as operated at predefined set points and therefore for example a consumer can not increase its power consumption, as the maximum power consumption is defined by the supply network temperature and the fixed mass flow passing through the heat exchanger of the consumer. Similarly, this limits the possible variation of heat power supply of the producers. This does not only prevent ideal coordination of supply and demand for example through DSM, but it also has a large impact on the technical efficiency of the DHN as explained in [MTK+22, pp. 6-7]. Heat losses mainly occur in DHNs, based on the heat flow through pipeline insulations. As heat losses increase with rising difference between the fluid and the ambient temperature, these can be minimized by keeping the temperatures as low as possible [ZLG13][19]. Therefore, DHN operators aim to set the mass flows through the heat exchangers of consumers and producers such that the temperature difference between supply and return network is maximized, and thereby the mass flows and the resulting pump costs are minimized [ZLG13]. As heat losses are in the range of 12 % to 20 % [VSLD13, ÇYÇ04, LMMD10], an efficient operation based on variable mass flows (and variable temperatures) is of significant interest.

[19] Supply network temperatures can only be reduced to a certain level in order to sufficiently supply the heat demand of customers.

- Stationary DHN pipeline models: As stated in [ZWW+21]: "Neglecting the dynamic characteristics of the DHS may result in an infeasible or suboptimal solution."[11] As will be demonstrated in Section 4.2.1 the storage potential in DHN pipelines can increase the flexibility of the entire CEPDHN and thus help to increase the social welfare as heat is fed into the DHN during time of low priced heat supply and then consumed in the following time steps with higher energy prices. Besides, this flexibility can help to prevent RESs curtailment [ZZZW18b], which also increases technical efficiency.
- Constant flow direction: Regarding Figure 1.1 using a model with fixed flow directions would prevent the ability of Producer 2 to provide Consumer 1 with heat power. Transferring this aspect to larger DHNs with more FNPs makes it apparent how severe the market outcome can be affected through this modeling decision. Besides, once more this model simplification can affect the technical efficiency, as due to this assumption, hot water might flow longer distances from producers farther away to specific consumers, creating higher heat losses.

None of the presented papers presented in the previous Sections 1.2.1 to 1.2.4, addressed the full necessary technical detail needed to achieve technical efficient and economic optimal CEPDHN operation. Thus, the second beneficial property, stated in Section 1.1, is so far not fulfilled within the state of the art.

Remark 1.2:
As discussed here, the optimization problem for market-based operation is either described by a significantly simplified convex model or by an accurate non-convex model. Based on the aforementioned, one cannot guarantee in practice finding the Pareto optimal solution for the real problem. When referring to ***economic optimal*** *solutions in this work, this always refers to a welfare maximizing solution of the given optimization problem. If this is obtained by a non-convex optimization problem, it should be kept in mind that this solution does not have to be the Pareto optimal solution, as it could represent a local optimum.*

Further, on the one side the more technical work presented in Section 1.2.1 and 1.2.2 lack the joint design of market and operational mechanisms, as the scope of these papers is not set towards economic issues. On the other side, most of the more economically oriented approaches presented in Section 1.2.3 and 1.2.4 cannot guarantee optimal spatial and temporal coordination of the FNPs as well as secure network operation[8], due to the strongly simplified network models. Therefore, the third beneficial property, deduced in Section 1.1, is not reached in the current literature. Still, market and control mechanisms have strong mutual interdependencies. Just to name a few, for example, pricing mechanisms will affect the Market Participants (MPs) bidding behavior, and thus the amount of power dispatchable at a certain point of time and a certain location. Further, defining prediction horizons and time step intervals in TC approaches, will directly impact the bid structures, the dispatch results and thereby also again the market prices. Also several interrelations of market and control mechanisms, explained in detail in Chapter 3, can on the one side limit the applicability of certain approaches, and on the other side increase the performance of the Transactive Control System (TCS) if adequately employed. Therefore, designing these market and control mechanisms in a joint coordinated process is a key element to achieve technical efficiency and economic optimality at the same time.

Distributed operation approaches, can help to overcome the computational burden of operational optimization of large CEPDHNs with numerous FNPs and hence enable scalability. A large amount of literature is found, proposing approaches for the distributed optimization of EPNs, see [MDS+17] for a review. However, so far the proposed distributed CEPDHN operation approaches, stated in Section 1.2.2, only decompose the CEPDHNs along EPN tie lines or the energy converters between DHNs and EPNs. The decomposition within DHNs is still to be performed. To the best of the authors knowledge the only approach provided so far is based on a MILP optimal DHN operation model [KMHG21]. Thus the authors of [ZWW+21] stated this as future work: "In solution methods for decentralized optimization, decomposition scheme suitable for more complex coupling constraints need to be exploited".[20] Note, that a clear motivation for distributed DHN operation approaches arises from the high computational cost of detailed DHN models [ZWW+21], as well as the simulation results from large real world DHNs, in Chapter 4.

To overcome the challenges stated in Section 1.1 it is of high interest that future CEPDHN operation approaches enable scalability, ensure secure network operation, coordinate FNPs in EPNs and DHNs spatially and temporally, prevent technical inefficiency, and provide market and control mechanisms that interact in a joint concerted form. Thus the research gap, defined by the deviation between these beneficial properties to the status quo and the current literature can be summarized as follows:

Research Gap:

1. *Market and operational control mechanisms have so far not been fully integrated for optimal coordination of FNPs in CEPDHNs.*
   a) *The necessary technical modeling detail for efficient CEPDHNs operation, is not reached within the state of the art on model-based CEPDHNs operation.*
   b) *The technical modeling detail directly impacts the market outcome. This aspect has not been sufficiently addressed within the state of the art on CEPDHN operation.*
2. *So far the proposed distributed CEPDHN operation approaches only decompose the CEPDHNs along EPN tie lines or the energy converters between DHNs and EPNs. The decomposition within DHNs is still to be performed.*

[20] Note, that the authors definition of *decentralized*, corresponds to the definition of *distributed* used within this work.

1.3 Research Objective and Statement of Contributions

In order to bridge the outlaid research gap, the objective and main contribution of this thesis is:

> **Contribution 1:**
> *The design of a new TCS for CEPDHNs. Thereby, bringing together market and control mechanisms in a form that enables a technically efficient and secure[8] operation for future CEPDHNs. Additionally, the proposed TCS enables parallelization of computations by applying a distributed optimization approach.*

The contributed approach is designed based on a TC framework as the aforementioned is best suited to provide the beneficial properties listed in Section 1.1. This is due to the fact, that distinctive features of TC approaches are the ability to coordinate supply and demand through market interactions, while providing control signals coordinating large numbers of FNPs to respect energy system operation constraints, enabling scalability and additionally providing salient features as privacy preserving information exchanges of smart energy management systems [KW16, LLJZ20]. Still, different TC approaches can vary strongly, see [LLJZ20]. Therefore, Contribution 1 comprises the definition of entities and their respective information exchange, the design of complementary market and control mechanisms, and the embedding in ambient control loops existing in EPNs and DHNs. As the chosen control mechanism is a distributed optimization based rolling horizon implementation, an important design criterion is the selection of an appropriate distributed optimization approach for NLP problems. Contribution 1 is presented in Chapter 3 and its application is demonstrated in Chapter 4. Note, that the central Contribution 1 is build upon two further contributions stated below, as depicted in Figure 1.2 showing the relation of the four contributions of this work. Within this TCS ap-

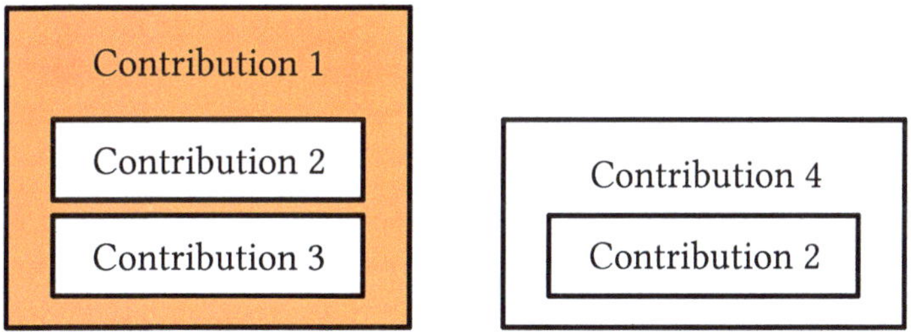

Figure 1.2: The relation of the four contributions of this thesis. Contribution 1 is built upon Contributions 2 and 3. Further, Contribution 4 is also based on Contribution 2.

proach, an adequate model of the CEPDHN is of high relevance in order to guarantee secure[8] and efficient CEPDHN operation based on calculations that can be performed at runtime.

Thus the second contribution of this thesis is:

> **Contribution 2:**
> *The composition of a novel detailed CEPDHN model for operational optimization. This model avoids the questionable simplifications discussed in Section 1.2.5.*

The tradeoff between computational burden and exactness of representation of physical effects lead to a nonlinear non-convex model, in which several variables are approximated by parameters which are adequately precalculated. While for several components of the CEPDHN the selection of an adequate model is performed by using models based on existing literature, a quantitative and qualitative comparison of existing dynamic thermal DHN pipeline models performed by the author in [MRMH21] showed that no DHN pipeline model fully suits the requirements. In this context, the main requirement is a high model accuracy as pipelines are the most common components and therefore imprecise models can lead to significant subsequent errors. Besides, the ability to take into account pipeline storage through dynamic models and the potential to incorporate VMFDs are important. Last but not least, the computational cost is to be kept limited. The newly developed dynamic DHN pipeline model, which is presented in Section 2.3.3 suits these requirements, and is designed to optimally fit within the control mechanisms. Furthermore, the overall CEPDHN model is able to take into account Variable Flow and Variable Temperature (VFVT) operation conditions with VMFDs, the dynamics of temperature front propagation, control of active hydraulic components as pumps, valves and DPRs, and the full AC power flow. Contribution 2 is presented in Chapter 2.

Also, different pricing mechanisms are compared for their suitability for coupled CEPDHN market clearing within the newly developed TCS. As no existing pricing mechanism is optimally suited for the given problem, a new pricing approach was developed, thus the last contribution is:

> **Contribution 3:**
> *A new hybrid pricing based mechanism for coordinated CEPDHN market clearing, in order to facilitate incentive compatibility while preventing uplift costs from redispatch.*

A detailed description of this approach is given in Section 3.2.3. Further, note that simulational results demonstrating the effectiveness of the stated contributions are presented in Chapter 4.

The last contribution of this work is:

> **Contribution 4:**
> *A distributed optimal operation approach for DHNs based on OCD.*

As described in detail in Section 3.3.2, OCD is well suited for the implementation of distributed optimal dispatch problems of EPNs. When applying OCD to DHNs, the symmetric structure of supply and return networks, and especially the representing network models are nontrivial to decompose into sub problems. Thus, the key challenge is to find a form of decomposition that reduces the coupling between the subproblems, in order to enable the distributed approach to converge to the same solution as a central optimization. At the same time, the proposed reformulation enables to take all physical constraints into account, while preventing the elimination of all degrees of freedom of the underlying models of the small subproblems. This was achieved by an appropriate reformulation of the initial optimal dispatch problem for DHNs, developed within Contribution 2. Based on the aforementioned, the application of OCD is demonstrated and the limitations of the approach are discussed. The methods are presented in Section 3.3.2.

1.4 Structure and Notation

In order to provide readability and orientation, the notation and structure of this work are provided below.

1.4.1 Structure

The introduction, stated above, is followed by the three main chapters. Chapter 2 presents the CEPDHN model, which comprises the EPN model, the EC models, and the thermal and hydraulic DHN models. All of these models, are optimally suited for utilization in the TCS system which is introduced in detail in Chapter 3. Therein the market and control mechanisms are designed in a coordinated form. This includes different pricing strategies, the control approaches, as well as presenting different distributed optimization algorithms. This is followed by a variety of case studies in Chapter 4, showing the practical usefulness of the newly developed TCS to different CEPDHNs, indicating various aspects of the market based operation as for example loss reduction, providing flexibility, computational burden and price signals arising from the selected pricing mechanism. Finally, the main conclusions of this work are drawn together in Chapter 5.

1.4.2 Notation

Matrices and vectors are written in bold symbols. For notational simplicity, the vertical concatenation of (row) vectors is written as e.g. $\boldsymbol{x}^{\mathrm{all}} = \left[\boldsymbol{x}^{\mathrm{za}}, \boldsymbol{x}^{\mathrm{zb}}\right]$ instead of the formal correct form $\boldsymbol{x}^{\mathrm{all}} = \left[(\boldsymbol{x}^{\mathrm{za}})^\top, (\boldsymbol{x}^{\mathrm{zb}})^\top\right]^\top$.

The gradient of function f of a stacked vector $\left[\boldsymbol{x}^{\mathrm{za}}, \boldsymbol{x}^{\mathrm{zb}}\right]$ is written as $\nabla f_{[\boldsymbol{x}^{\mathrm{za}},\boldsymbol{x}^{\mathrm{zb}}]}$, while the second-order derivative of f of $\boldsymbol{x}^{\mathrm{za}}$ and $\boldsymbol{x}^{\mathrm{zb}}$ is written as $\nabla^2 f_{\boldsymbol{x}^{\mathrm{za}},\boldsymbol{x}^{\mathrm{zb}}} = \nabla_{\boldsymbol{x}^{\mathrm{zb}}}(\nabla_{\boldsymbol{x}^{\mathrm{za}}} f)$.

If not further clarified, all variables and parameters are real-valued. Complex symbols are written in underlined form, e.g. the admittance $\underline{Y}$. Indices and superscripts written in italic form represent variable entries, while non italic superscripts represent identifiers.

2 Modeling of Coupled Electric Power and District Heating Networks

Representing CEPDHNs in adequate models for operational optimization is a complex task, as the characteristics of CEPDHNs bring along [ZWW+21]: "a large computational burden and nonlinear and non-convex properties." Thus, the combination of suitable component models, representing the relevant physical effects while preserving real time computation is of central importance for this work. After an introduction, focusing on the main physical effects in CEPDHNs and the properties of the EPN, the DHNs, and the Network Participants (NPs), the single components will be presented. Note, that NPs include producers, consumers, storages and energy converters between the EPN and the DHNs. All together, these models form a new detailed CEPDHN model for the TCS presented in this work.

The following Remark 2.1 helps to understand the aim of the model design and the terminology used within this chapter:

Remark 2.1:
The TCS design is based on a model-based rolling horizon[21] *approach. The basic idea of this control mechanism is to carry out an optimization, based on a dynamic system model, of the CEPDHNs, to determine the optimal control values. The optimization problem solved for every time step k takes into account all time steps on the* ***rolling horizon*** *$k \in \mathbb{S}_k$. Then the control values of the last optimization are sent to the controlled units, measurement data is used to update the system state variables, the rolling horizon time shift is performed and the optimization is carried out again. This procedure is continuously repeated during system operation. Since the presented model can also be used to simulate CEPDHNs, when the control values are entered as predetermined parameters, the term* ***simulation horizon*** *will be used in the same sense as prediction horizon in the following descriptions.*

The presented overall CEPDHN model enables efficient operation through high technical detail, while maintaining computability, forming Contribution 2 of this thesis. Parts of this chapter have been published in [MEKH18, MRMH21], and [MTK+22].

Within this work, the CEPDHN is modeled as a combination of multiple directed graphs. Therein, the EPN and each DHN are modeled as a directed graph. These initially separated graphs are coupled through ECs, which represent a subset of the NPs, connecting buses of the EPN and respective edges in the DHNs, see Figure 2.1. The DHN network graph elements, are defined by the set of all edges $e \in \mathbb{S}_e^{\text{dhn}}$ and all nodes $i \in \mathbb{S}_i^{\text{dhn}}$. Due to the interplay of supply and return networks, these DHN graphs are strongly connected[22]. The connectivity

[21] Also known as receding or moving horizon [Arn11].
[22] If only the supply network is regarded, the directed graph becomes weakly connected.

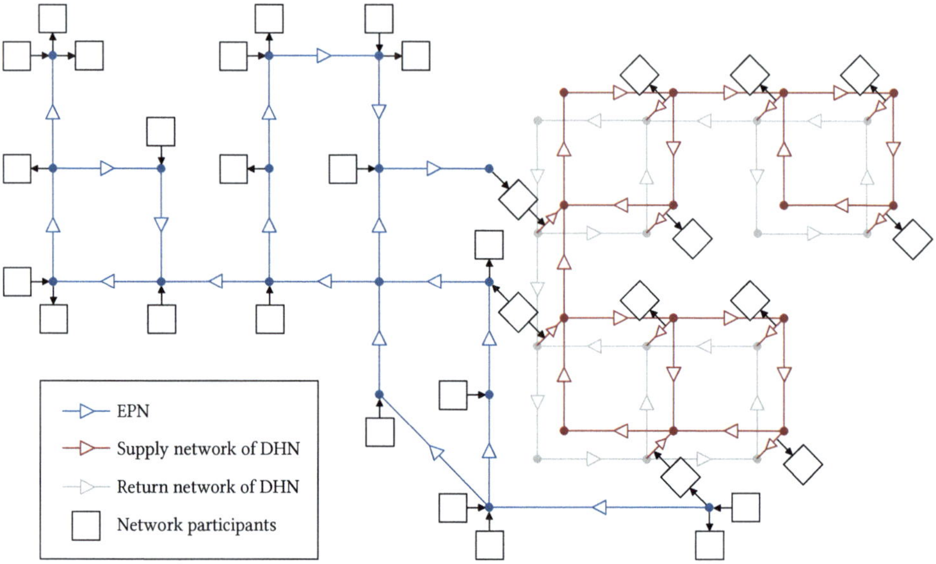

Figure 2.1: Exemplary network of the entire CEPDHN structure.

of nodes i and edges e within a DHN is defined within the DHN node edge incidence matrix $\boldsymbol{A}^{\mathrm{dhn}}$, defined as[23]:

$$A_{i,e}^{\mathrm{dhn}} = \left\{ \begin{array}{r} 1, \text{ if } e \text{ is directed towards } i, \\ -1, \text{ if } e \text{ is directed away from } i, \\ 0, \text{ if } e \text{ is not connected to } i, \end{array} \right\} \forall\, i \in \mathbb{S}_i^{\mathrm{dhn}}, e \in \mathbb{S}_e^{\mathrm{dhn}} \tag{2.1}$$

Note, that the EPN node edge incidence matrix $\boldsymbol{A}^{\mathrm{epn}}$ is modeled similarly, by exchanging the respective sets in equation (2.1) to the set of all buses of the EPN $\mathbb{S}_i^{\mathrm{epn}}$ and the set of all edges of the EPN, specifically $\mathbb{S}_e^{\mathrm{epn}}$. The EPN graph spanned up by $\boldsymbol{A}^{\mathrm{epn}}$ is assumed to be a weakly connected graph, thereby the case of multiple unconnected power grids is neglected.[24]

[23] For notational simplicity and without the loss of generality, the DHN model given within this chapter is limited to a specific DHN, leaving aside ability to distinguish between multiple DHNs.

[24] This does not represent a major limitation of the approach as islanded micro grids represent an absolute minor part in the German and central European power supply, which is considered as the focus of this work.

Table 2.1: Signal propagation compared for different physical values relevant in CEPDHN models [Ill20, p. 15], [SGS+17, p. 3015], and [Opp15, p. 25].

Network type	**EPN**	**DHN** Hydraulic	**DHN** Thermal
Potential value	Voltage	Pressure	Temperature
Propagation speed	Speed of light ($\sim 1.8 \times 10^8\,\mathrm{m/s}$)	Speed of sound ($\sim 1000\,\mathrm{m/s}$ to $1500\,\mathrm{m/s}$)	Flow velocity ($\sim 0\,\mathrm{m/s}$ to $3\,\mathrm{m/s}$)
Example distance		$\sim 1000\,\mathrm{m}$	
Example propagation time	$\sim 3.33\,\mu\mathrm{s}$	$\sim 0.7\,\mathrm{s}$ to $1\,\mathrm{s}$	$\sim 333.33\mathrm{s}$ to $\infty\mathrm{s}$

The connections of NPs, which are either linked to edges in the DHNs, buses in the EPN or to both if they represent energy converters, are described in the bus-NP incidence matrix[25,26], for the EPN $\boldsymbol{M}^{\mathrm{epn}}$ and the edge-NP incidence matrix for the DHN $\boldsymbol{M}^{\mathrm{dhn}}$:

$$M_{i,n}^{\mathrm{epn}} = \begin{cases} 1, & \text{if } n \text{ is a network participant connected to } i, \\ 0, & \text{else,} \end{cases} \quad \forall\, i \in \mathbb{S}_i^{\mathrm{epn}}, n \in \mathbb{S}_n \tag{2.2}$$

$$M_{e,n}^{\mathrm{dhn}} = \begin{cases} 1, & \text{if } n \text{ is a network participant connected to } e, \\ 0, & \text{else,} \end{cases} \quad \forall\, e \in \mathbb{S}_e^{\mathrm{dhn}}, n \in \mathbb{S}_n \tag{2.3}$$

Therein, all network participants n are accumulated in $\mathbb{S}_n$. Based on the above mentioned, the entire CEPDHN structure is defined as shown in Figure 2.1.

An important physical aspect for CEPDHN models for operational optimization is the large difference in signal propagation of voltages, pressures and temperatures as stated in Table 2.1: Typical time step intervals of operational optimization of EPNs and DHNs , as economic dispatch with elastic demand, are within the scale of several minutes [LGZ+19]. By taking into account the propagation times from Table 2.1, it can be deduced that the EPN and the hydraulic effects of the DHN can be modeled in stationary state[27], while the propagation of thermal fronts needs to be represented by dynamic models, to prevent suboptimal or infeasible solutions within operational optimization approaches. This is a common assumption used in modeling of DHNs [ZWW+21]. A further important difference between DHN and EPN models, that is given in Table 2.1 is the fact, that DHN models possess two effort variables, pressure and temperature, while in contrast these are only represented by voltages in EPNs. Of course, apart from the mentioned differences, several obvious parallels exist between the stationary hydraulic DHN and the EPN models, as e.g. the Kirchhoffs circuit laws are similarly repre-

[25] Note, that in general NPs could also be modeled as connected to nodes in the DHNs. However in this case the return network and the DHN components between the return and the supply network as e.g. heat exchangers, pumps, and valves, would have to be neglected.

[26] Note that power flows from the DHN to the EPN are not precluded by the modeling approach, but do not play a relevant role in the practical cases regarded here.

[27] Since e.g. dynamic frequency or pressure regulation are not performed in this time scale and the respective transient effects are thus neglectable provided adequate controls for the before mentioned.

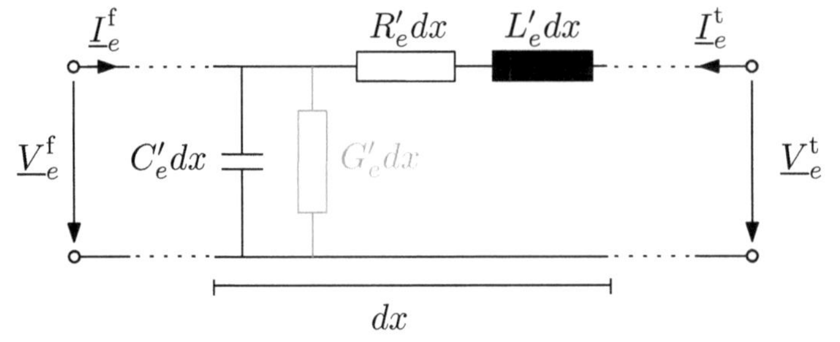

Figure 2.2: Infinitesimal element of length dx of transmission line of network edge e [GECC18, p. 68]. The superscripts f and t of voltages V and currents I label the two ends of the transmission line, representing *from* and *to*. For usual power lines, the shunt conductance G' is neglectable [Kun93, p. 204], which is why it is only indicated in this Figure.

sented within the hydraulic models, as depicted in the following of this chapter. All model components of the CEPDHN model for operational optimization are given in detail below.

2.1 Electric Power Network

This section describes the EPN model used within the TCS for CEPDHNs. The modeling of power flows in EPNs is a well studied field. Several well known models exist, which represent the power flows over transmission lines between network buses. These models differ in DC vs AC, transmission vs distribution, convex vs non-convex, linear vs nonlinear, limited to radial vs applicable to meshed network structure, and lossy vs lossless network representation, see e.g. [MDS$^+$17]. Since the energy converters, see Section 2.2, are connected to electric distribution networks [CWL$^+$18], this work focuses on electric distribution systems[28] within the CEPDHNs. The subsequent Sections 2.1.1 until 2.1.3 present the used transmission line, bus, and overall network model.

2.1.1 Transmission Line

Assuming balanced and sinusoidal steady-state conditions[29], electric transmission lines, can be modeled as an infinite sequence of infinitesimal elements of length dx, as illustrated in Figure 2.2 [GECC18, p. 68]: where the parameters of the series resistance R'[30], the series inductance $L^{\text{ind}\prime}$, the shunt conductance G', and the shunt capacitance C' are all defined per length and per phase.

---

[28] Leaving electric transmission networks aside.

[29] The later can be derived from Table 2.1.

[30] Note, that indices are only provided with the respective symbols in the text, if they are of relevance within the explanation.

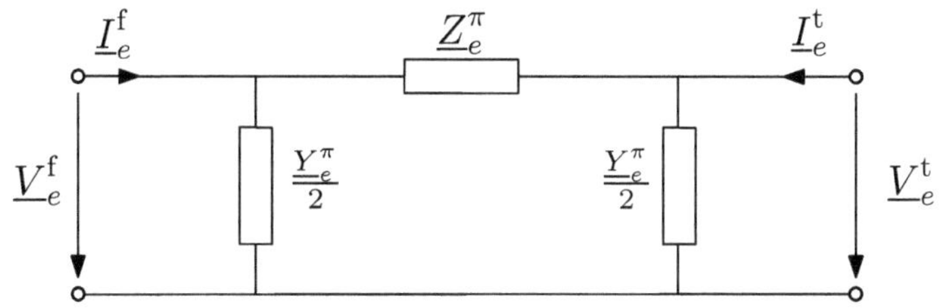

Figure 2.3: π-equivalent circuit of a transmission line using lumped parameters, see e.g. [Kun93, p. 207].

Based on the parameters, the serial impedance $\underline{Z}'$ and shunt admittance $\underline{Y}'$ per length and phase for every edge e of the set of all EPN edges $\mathbb{S}_e^{\mathrm{epn}}$ are defined by [Kun93, p. 202]:

$$\underline{Z}'_e = R'_e + j\omega {L^{\mathrm{ind}}}'_e, \quad \underline{Y}'_e = G'_e + j\omega C'_e, \quad \forall\, e \in \mathbb{S}_e^{\mathrm{epn}} \tag{2.4}$$

Based on the aforementioned, the characteristic impedance $\underline{Z}^{\mathrm{char}}$ and the propagation constant $\underline{\gamma}^{\mathrm{pc}}$ are calculated by[MBB11, p. 66]:

$$\underline{Z}_e^{\mathrm{char}} = \sqrt{\frac{\underline{Z}'_e}{\underline{Y}'_e}}, \quad \underline{\gamma}_e^{\mathrm{pc}} = \sqrt{\underline{Z}'_e \underline{Y}'_e}, \quad \forall\, e \in \mathbb{S}_e^{\mathrm{epn}} \tag{2.5}$$

Using lumped parameters for a given transmission line e of length L_e, the simplified π-equivalent circuit, as shown in Figure 2.3 can be stated, with the π-equivalent impedance $\underline{Z}_e^{\pi}$ and admittance $\underline{Y}_e^{\pi}$ given by [Kun93, p. 207]:

$$\underline{Z}_e^{\pi} = \underline{Z}_e^{\mathrm{char}} \sinh\left(\underline{\gamma}_e^{\mathrm{pc}} L_e\right), \quad \forall\, e \in \mathbb{S}_e^{\mathrm{epn}} \tag{2.6}$$

$$\frac{\underline{Y}_e^{\pi}}{2} = \frac{1}{\underline{Z}_e^{\mathrm{char}}} \tanh\left(\frac{\underline{\gamma}_e^{\mathrm{pc}} L_e}{2}\right), \quad \forall\, e \in \mathbb{S}_e^{\mathrm{epn}} \tag{2.7}$$

Based on the aforementioned, the transmission line admittance matrix $\underline{\boldsymbol{Y}}_e^{\mathrm{trl}}$ of network edge e can be calculated by:

$$\underline{\boldsymbol{Y}}_e^{\mathrm{trl}} = \begin{bmatrix} \left.\frac{\underline{I}_e^{\mathrm{f}}}{\underline{V}_e^{\mathrm{f}}}\right|_{\underline{V}_e^{\mathrm{t}}=0} & \left.\frac{\underline{I}_e^{\mathrm{f}}}{\underline{V}_e^{\mathrm{t}}}\right|_{\underline{V}_e^{\mathrm{f}}=0} \\ \left.\frac{\underline{I}_e^{\mathrm{t}}}{\underline{V}_e^{\mathrm{f}}}\right|_{\underline{V}_e^{\mathrm{t}}=0} & \left.\frac{\underline{I}_e^{\mathrm{t}}}{\underline{V}_e^{\mathrm{t}}}\right|_{\underline{V}_e^{\mathrm{f}}=0} \end{bmatrix} = \begin{bmatrix} \frac{1}{\underline{Z}_e^{\pi}} + \frac{\underline{Y}_e^{\pi}}{2} & -\frac{1}{\underline{Z}_e^{\pi}} \\ -\frac{1}{\underline{Z}_e^{\pi}} & \frac{1}{\underline{Z}_e^{\pi}} + \frac{\underline{Y}_e^{\pi}}{2} \end{bmatrix}, \quad \forall\, e \in \mathbb{S}_e^{\mathrm{epn}} \tag{2.8}$$

This transmission line admittance matrix is then used in Section 2.1.3 to set up the nodal admittance matrix $\underline{\boldsymbol{Y}}^{\mathrm{bus}}$ describing the admittances of the entire EPN.

2.1.2 Bus

Shunt admittances represent the connection of a bus i to the ground:

$$\underline{Y}_i^{\text{sh}} = G_i^{\text{sh}} + jB_i^{\text{sh}}, \quad \forall\, i \in \mathbb{S}_i^{\text{epn}} \tag{2.9}$$

Concatenating all shunts of the EPN into the column vector $\underline{\boldsymbol{Y}}^{\text{sh}}$ is written as:

$$\underline{\boldsymbol{Y}}^{\text{sh}} = \left[\underline{Y}_1^{\text{sh}}, \ldots, \underline{Y}_i^{\text{sh}}, \ldots, \underline{Y}_{\text{n}^{\text{ci}}}^{\text{sh}}\right] \tag{2.10}$$

with the cardinality of the EPN buses given by $\text{n}^{\text{ci}} = |\mathbb{S}_i^{\text{epn}}|$. Also, this vector will be used in the following Section 2.1.3 to describe the entire EPN.

The real and reactive power, $P_{i,k}$ and $Q_{i,k}$, injected at bus i and time step k by all network participants n connected at this bus, is defined as:

$$P_{i,k} = \sum_{n \in \mathbb{S}_n^{\text{epn}}} M_{i,n}^{\text{epn}} P_{n,k}, \qquad \forall\, i \in \mathbb{S}_i^{\text{epn}},\ k \in \mathbb{S}_k \tag{2.11}$$

$$Q_{i,k} = \sum_{n \in \mathbb{S}_n^{\text{epn}}} M_{i,n}^{\text{epn}} Q_{n,k}, \qquad \forall\, i \in \mathbb{S}_i^{\text{epn}},\ k \in \mathbb{S}_k \tag{2.12}$$

where the elements of the bus-NP incidence matrix $M_{i,n}^{\text{epn}}$ are defined as in (2.2) and $\mathbb{S}_k$ comprises all time steps k on the prediction/simulation horizon[31].

2.1.3 Network

The nodal admittance matrix $\underline{\boldsymbol{Y}}^{\text{bus}}$ of the entire EPN is calculated by [ZMS11]:

$$\underline{\boldsymbol{Y}}^{\text{bus}} = \underline{\boldsymbol{A}}^{\text{epn},-}\underline{\boldsymbol{Y}}^{\text{f}} + \underline{\boldsymbol{A}}^{\text{epn},+}\underline{\boldsymbol{Y}}^{\text{t}}\text{diag}\left(\underline{\boldsymbol{Y}}^{\text{sh}}\right) \tag{2.13}$$

where the auxiliary admittance matrices $\underline{\boldsymbol{Y}}^{\text{f}}$ and $\underline{\boldsymbol{Y}}^{\text{t}}$ are built by parameters from the transmission line admittance matrices $\underline{\boldsymbol{Y}}_e^{\text{trl}}$ of the network edges as given in the Appendix A. Further, the incoming and leaving edge bus incidence matrices $\underline{\boldsymbol{A}}^{\text{epn},+}$ and $\underline{\boldsymbol{A}}^{\text{epn},-}$ are defined as:

$$A_{i,e}^{\text{epn},+} = \begin{Bmatrix} 1, \text{ if } A_{i,e}^{\text{epn}} = +1 \\ 0, \text{ else}, \end{Bmatrix} \qquad \forall\, i \in \mathbb{S}_i^{\text{epn}}, e \in \mathbb{S}_e^{\text{epn}}$$

$$A_{i,e}^{\text{epn},-} = \begin{Bmatrix} 1, \text{ if } A_{i,e}^{\text{epn}} = -1 \\ 0, \text{ else}, \end{Bmatrix} \qquad \forall\, i \in \mathbb{S}_i^{\text{epn}}, e \in \mathbb{S}_e^{\text{epn}} \tag{2.14}$$

Note, that $A_{i,e}^{\text{epn}}$ is defined analog to $A_{i,e}^{\text{dhn}}$ in (2.1). The following property of the nodal admittance matrix $\underline{\boldsymbol{Y}}^{\text{bus}}$ will be considered in Section 3.3.2:

31 See Definition 2.3 for more information.

Remark 2.2:
The nodal admittance matrix $\underline{\mathbf{Y}}^{\text{bus}}$ is typically sparse, and the degree of sparsity increases with the size of the EPN network size [Kun93, p. 258].

An accurate model of the meshed lossy AC distribution system is based on the nonlinear and non-convex[32] AC power flow equations [MBB11, p. 115]:

$$P_{i,k} = \sum_{j \in \mathbb{S}_i^{\text{epn}}} V_{i,k} V_{j,k} [B_{i,j}^{\text{bus}} \sin(\delta_{i,k} - \delta_{j,k}) + G_{i,j}^{\text{bus}} \cos(\delta_{i,k} - \delta_{j,k})], \qquad \forall\, i \in \mathbb{S}_i^{\text{epn}},\ k \in \mathbb{S}_k \tag{2.15}$$

$$Q_{i,k} = \sum_{j \in \mathbb{S}_i^{\text{epn}}} V_{i,k} V_{j,k} [G_{i,j}^{\text{bus}} \sin(\delta_{i,k} - \delta_{j,k}) - B_{i,j}^{\text{bus}} \cos(\delta_{i,k} - \delta_{j,k})], \qquad \forall\, i \in \mathbb{S}_i^{\text{epn}},\ k \in \mathbb{S}_k \tag{2.16}$$

where $P_{i,k}$ and $Q_{i,k}$ denote the active and reactive power injected into the system at bus i at time k by all network participants n connected at this bus, as defined in equation (2.11) and (2.12). The conductance $G_{i,j}^{\text{bus}}$ and the susceptance $B_{i,j}^{\text{bus}}$ refer to the nodal admittance matrix element $\underline{Y}_{i,j}^{\text{bus}}$. The bus voltage angles are given as δ.

2.2 Energy Converter

The energy converters, such as heat pumps, electric boilers or CHP units are often modeled by constant (efficiency) coefficients in most MES models for operational optimization, see e.g. [GA07]. A more detailed model, which incorporates this constant (efficiency) coefficient model form, is given using polynomial functions, representing the Coefficient of Performance (COP) or rather the coupling factor in general, as follows [Sau19, p. 108]:

$$\Phi_{n^{\text{dhn}},k} = \sum_{m=0}^{3} \sum_{n^{\text{epn}} \in \mathbb{S}_n^{\text{epn,ec}}} a_{n^{\text{epn}},n^{\text{dhn}},m} \cdot \left(P_{n^{\text{epn}},k}\right)^m, \quad \forall\, n^{\text{dhn}} \in \mathbb{S}_n^{\text{dhn,ec}},\ k \in \mathbb{S}_k \tag{2.17}$$

Therein, the (polynomial) coefficients a are $\in \mathbb{R}$, the polynomial order is given by m, the heat power output of an EC is defined by Φ, and the electric real power injected (or consumed) is stated as P. Further, the index n representing an FNP is provided here with an EPN or DHN superscript to specify the connected network and the set of all network participants n representing energy converters connected to the DHN is given by $\mathbb{S}_n^{\text{dhn,ec}}$. Note, that the presented model is not limited to these kind of polynomial energy converter models. More complex functions of the coupling factor of energy converters in MESs, as e.g. presented in [MSKH16] can also be included. Further, as an alternative to (2.17) descriptions using possible areas of operation through polygons, see Figure 2.4, which are then described by inequalities,

[32] The nonlinear and non-convex form of these equations has not prevented numerous successful implementations, e.g. Optimal Power Flow (OPF) problem based applications, using this exact model form [MDS+17].

limiting possible ratios of Φ to P instead of equation (2.17) can be used [DLS+19, LWZS17]:

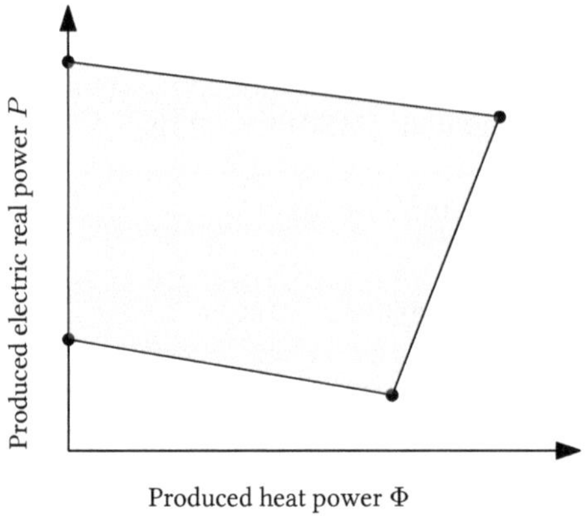

Figure 2.4: Feasible region of the operation ratio of $\Phi_{n,k}$ to $P_{n,k}$ of an extraction-condensing CHP unit n.

$$a_{1,b,n} P_{n^{\mathrm{epn}},k} + a_{2,b,n} \Phi_{n^{\mathrm{dhn}},k} \leq a_{3,b,n}, \qquad \forall\, n^{\mathrm{dhn}} \in \mathbb{S}_n^{\mathrm{dhn,ec}},\ b \in \mathbb{S}_b,\ k \in \mathbb{S}_k \tag{2.18}$$

With the coefficients a_1, a_2 and a_3 given for every energy converter modeled by this set of inequalities and the set $\mathbb{S}_b$ is comprised of all boundaries b. These inequalities (2.18) are often used to model the feasible operation areas of CHPs [ZWW+21].

2.3 District Heating Network

The most relevant components considered within operational optimization of DHNs are pumps, pipes, valves, DPRs, storages, heat exchangers, and the connectors of pipelines, typically referred to as nodes. Note, that producers, energy converters and consumers are only modeled until the heat exchanger side, directly connected to the DHNs, as usually done within the state of the art [ZWW+21]. As warm water is transported from producers to consumers through the supply network, and back through the return network, these networks are inherently meshed. The form of implemented network operation influences the network model, thus the typical forms of DHN operation [ZZZW18a] are listed below:

- Constant Flow Variable Temperature (CFVT) operation: Thereby, the mass flows through the edges are kept constant, while the temperatures, especially of the water leaving the producers, is varied.
- Variable Flow Constant Temperature (VFCT) operation: Therein, the mass flows are varied to meet changing energy demand of the DHN consumers, while the temperature supplied by the producers is kept constant.

- Variable Flow Variable Temperature (VFVT) operation: Here, mass flows and temperatures through the network are varied in order to keep losses as small as possible. In this form of operation, the set points of valves, pumps and the power supplied to heat exchangers within the producers is varied in a coordinated form.

Note, that within the CFVT and the VFCT operation modes the constant values, mass flows or temperatures, can be varied step wise, e.g. dependent on the ambient temperature T^{a}. Also, many DHNs are operated based on a combination of weather forecast, historical measurement data, as well as expert knowledge and experience of operation staff. However, it is well known that using the full system flexibility by the VFVT operation mode brings the highest efficiency [DLS+19, ZZZW18a][33]. Therefore, this form of network operation is envisioned, for the designed model, in this thesis. As the CFVT and the VFCT operation modes are a special case of the VFVT network operation, a model designed for this form of operation can also be applied in case a DHN is operated with CFVT or VFCT conditions. Therefore, this selection of operation form for the DHN model does not represent a limitation but instead enables to represent all major forms of network operation.

The characteristics of the most important elements in DHNs are described in the following. These are nodes, loops, control paths, pumps, DPRs, valves, heat exchangers, pipelines, and TESSs. Therefore, hydraulic and the thermal models of the network components are given. The hydraulic model defines the mass flows and differential pressures throughout the DHN. The thermal model describes the temperature changes in the different network components and the propagation of temperature fronts throughout the network. Before the hydraulic and the thermal models are presented, the most relevant material properties of water are given below.

2.3.1 Material Properties of Water

The density ρ^{w}, the specific heat capacity c^{w} and the kinematic viscosity[34] ν of liquid water are dependent on temperature and pressure levels. In Figure 2.5, the degree of these variations is depicted for possible ranges of pressure and temperature values in DHN operation. Therefrom, it can be seen that the density reaches values of $939.16\,\mathrm{kg/m^3} \leq \rho^{\mathrm{w}} \leq 998.35\,\mathrm{kg/m^3}$. Thus, modeling water as incompressible, utilizing a value of $\rho^{\mathrm{w}} = 960\,\mathrm{kg/m^3}$, leads to a resulting maximal possible deviation of $3.99\,\%$. Similarly, the specific heat capacity ranges from $4.173\,\mathrm{kJ/(kg\,K)} \leq c^{\mathrm{w}} \leq 4.255\,\mathrm{kJ/(kg\,K)}$. With a constant value of $c^{\mathrm{w}} = 4.214\,\mathrm{kJ/(kg\,K)}$ the largest occurring error is $0.97\,\%$. Hence, the resulting errors, brought into the model by constant values of ρ^{w} and c^{w}, are assumed to be negligibly small as they are below $4\,\%$.

More detailed examination is needed for the kinematic viscosity as value ranges are within $0.2366 \times 10^{-6}\,\mathrm{m^2/s} \leq \nu \leq 1.8 \times 10^{-6}\,\mathrm{m^2/s}$. Consequently, a kinematic viscosity of $\nu = 0.8369 \times 10^{-6}\,\mathrm{m^2/s}$ leads to a possible deviation of $115.08\,\%$. To understand the impact of this deviation on the pressure difference over a pipeline a parameter study was performed, which

[33] In this context, it should also be noted, that within the aforementioned forms of operation, different forms of pump operation can be found. A good overview on these is provided in [NTJK20].

[34] Note, that there is a difference between the dynamic viscosity and the kinematic viscosity which are both utilized throughout the literature, typically as constant values e.g. in [Trö99].

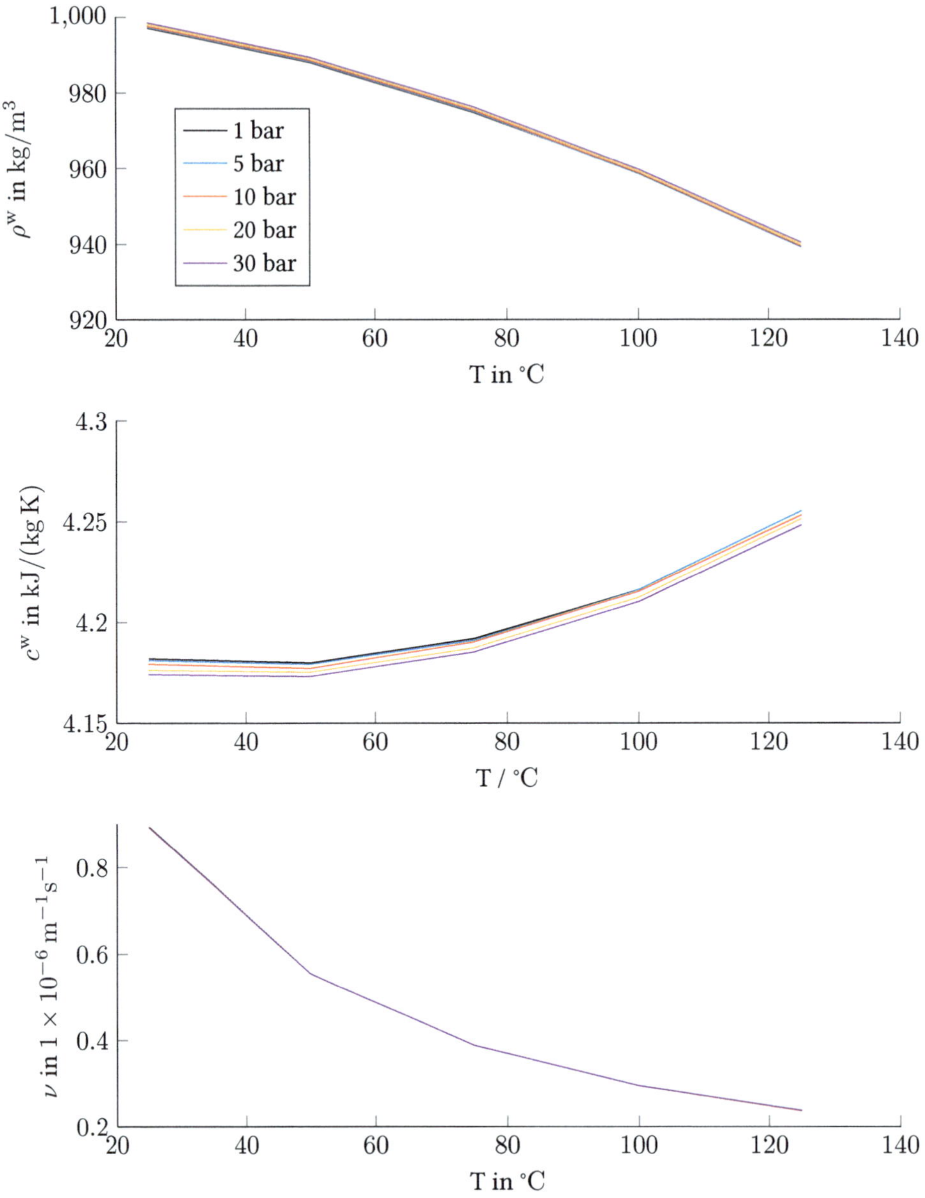

Figure 2.5: Values of the density ρ^{w}, the specific heat capacity c^{w} and the kinematic viscosity ν of liquid water for different temperatures and absolute pressure levels from [Els18, VDI13]. These are possible operation values of DHNs, see e.g. [ZWW+21, NTJK20]. The values of the kinematic viscosity change strongly for temperatures $\geq 100\,^{\circ}\mathrm{C}$ combined with low pressure values of 1 bar [VDI13]. These conditions are avoided in DHN operation, by setting the minimal absolute pressure $p^{\min}$ accordingly > 1 bar, to prevent evaporation [NTJK20] and are thus not plotted here for the kinematic viscosity. Exemplary values of e.g. $p^{\min} = 3$ bar are given in [NTJK20, p. 57].

is shown in Figure 2.6. Therein, the pressure difference over a pipeline is given for possible flow velocities v and the minimal, maximal and utilized value of the kinematic viscosity ν. The maximal possible deviation between these differential pressures calculated with the minimal or maximal and the utilized value of ν is shown in red. Similarly, as with the approximation of the friction factor calculation in Section 2.3.2, the possible deviation reaches values of $35.85\,\%$ for less likely low flow velocities and drops below $15\,\%$ for typical flows of $v \geq 1.4\,\mathrm{m/s}$. For the proposed rolling horizon approach, see Remark 2.1, the obtained accuracy is seen as sufficient and thus this aspect is disregarded. However, if more accuracy is needed it is possible to integrate look-up tables into the implementation which adapt the values of $\nu_{e,k}$ for every pipeline and time step before the optimization.

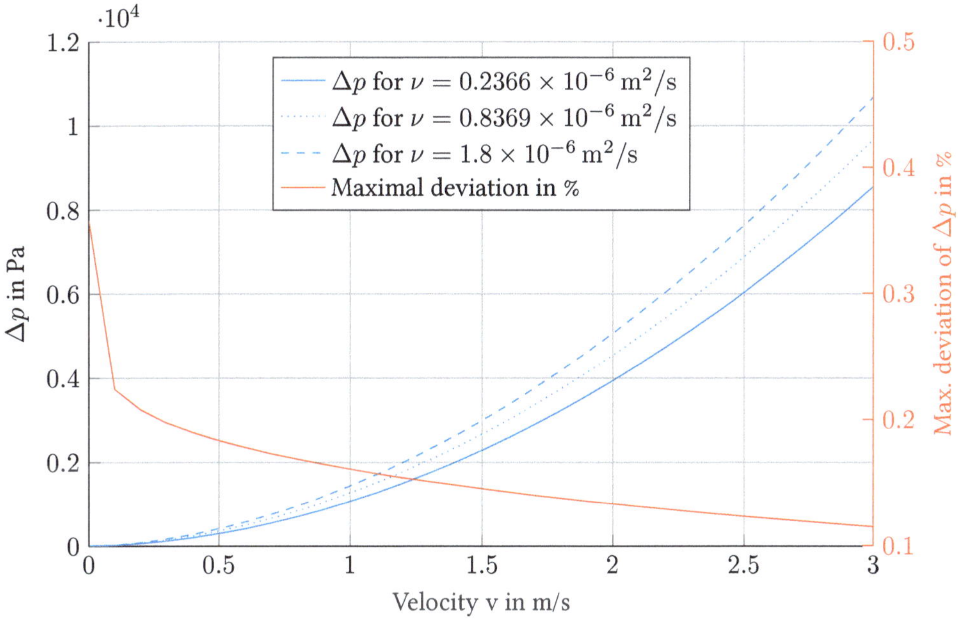

Figure 2.6: Influence of varying kinematic viscosity ν on the resulting differential pressure over a pipeline, based on the parameters for a pipeline taken from the hydraulic pipeline calculations in Section 2.3.2.

Summarizing the aforementioned leads to:

Assumption 2.1. *The resulting errors, brought into the model by constant values of the density* ρ^{w}*, the specific heat capacity* c^{w}*, and the kinematic viscosity* ν *of water are negligibly small.*

Besides the aforementioned, a typical assumption usually used within operational optimization models, and also within most DHN simulation models is: [OUGP16, DAM19]:

Assumption 2.2. *The cross-section of a component, e.g. a pipeline, has spatially homogeneous velocity and temperature throughout.*

Using this assumptions leads to much simpler 1D component models.

2.3.2 Hydraulic model

The models of all typical hydraulic elements are given below. For all components representing DHN edges e, a general stationary relation of the mass flow $\dot{m}$ and the pressure difference can be defined as [Ill20]:

$$\Delta p_{e,k} = \beta_{e,k} + \mu_{e,k}\dot{m}_{e,k}|\dot{m}_{e,k}|_{\Delta\varepsilon}, \qquad \forall\, e \in \mathbb{S}_e^{\mathrm{dhn}},\ k \in \mathbb{S}_k, \tag{2.19}$$

where β and μ are the variable component coefficients, which are given in detail for each component below. This equation is continuously differentiable as the absolute value function of the mass flow $\dot{m}$ is replaced by a continuous differentiable approximation given by:

$$|\dot{m}_{e,k}|_{\Delta\varepsilon} = \sqrt{\Delta\varepsilon + (\dot{m}_{e,k})^2} \tag{2.20}$$

wherein $\Delta\varepsilon > 0$ is a sufficiently small parameter.

In case of predefined positive flow direction[35], thus when the e is not element of the set of all edges with VMFDs $\mathbb{S}_e^{\mathrm{vmfd}}$, (2.19) can be written as:

$$\Delta p_{e,k} = \beta_{e,k} + \mu_{e,k}(\dot{m}_{e,k})^2, \qquad \forall\, e \in \mathbb{S}_e^{\mathrm{dhn}}, e \notin \mathbb{S}_e^{\mathrm{vmfd}},\ k \in \mathbb{S}_k \tag{2.21}$$

The pressure difference Δp over the edges is defined as:

$$\Delta p_{e,k} = -\sum_{i \in \mathbb{S}_i^{dhn}} A_{i,e}^{\mathrm{dhn}} p_{i,k}, \qquad \forall\, e \in \mathbb{S}_e^{\mathrm{dhn}},\ k \in \mathbb{S}_k \tag{2.22}$$

where p_i describes the absolute nodal pressure values[36].

Node The law of conservation of mass is represented by (2.23)[37] to model the equality of mass flows entering and leaving a node i:

$$\sum_{e \in \mathbb{S}_e^{\mathrm{dhn}}} A_{i,e}^{\mathrm{dhn}} \dot{m}_{e,k} = 0, \qquad \forall\, i \in \mathbb{S}_i^{\mathrm{dhn}},\ k \in \mathbb{S}_k \tag{2.23}$$

Therein, the mass flow through edge e at time step k is defined as $\dot{m}_{e,k}$ and the DHN node edge incidence matrix elements $A_{i,e}^{\mathrm{dhn}}$ are defined as in (2.1).

Loop The second important conservation law used to model the elementary hydraulic processes is the conservation of energy. This states, that the pressure differences $\Delta p_{e,k}$ along a

35 Flow direction equals the edge orientation.

36 A pressure maintenance unit guarantees a constant base pressure p_0 at a distinct node of the DHN This is typically a node in the return network near larger DHN pumps [NTJK18]. Based on this, the pressure differences created through water loss and density variations of the medium caused by temperature variations are balanced.

37 When only the supply network is modeled for simplicity reasons, the right hand side of (2.23) is extended by nodal mass flows, describing the mass flows fed into the supply network by e.g. producers or storages [MGR+21].

loop l must add up to zero:

$$\sum_{e \in \mathbb{S}_e^{\text{dhn}}} B_{l,e} \Delta p_{e,k} = 0, \qquad \forall\, l \in \mathbb{S}_l^{\text{dhn}},\ k \in \mathbb{S}_k \tag{2.24}$$

where the set of all loops of the DHN is $\mathbb{S}_l^{\text{dhn}}$ and the loop edge incidence matrix elements are given as:

$$B_{l,e} = \left\{ \begin{array}{r} 1, \text{ if } e \text{ is directed as } l, \\ -1, \text{ if } e \text{ is directed in opposition to } l, \\ 0, \text{ if } e \text{ is not element of } l, \end{array} \right\} \ \forall\, l \in \mathbb{S}_l^{\text{dhn}}, e \in \mathbb{S}_e^{\text{dhn}} \tag{2.25}$$

Control Path Besides the loops, control paths are also not component models, as they model the interaction of multiple components. Control paths consider the influence of controlled pumps and valves on the respective network edges, from the controlled element to the element where the controlled variable is measured. Typically the controlled variable is the predefined differential pressure Δp^{pre} over a pump, a set of edges, a valve or a consumer substation. The control path is then modeled, by defining, that the pressure drop over all edges of the control path should equal Δp^{pre}:

$$\sum_{e \in \mathbb{S}_e^{\text{dhn}}} R_{cp,e} \Delta p_{e,k} = \Delta p^{\text{pre}}, \qquad \forall\, cp \in \mathbb{S}_{cp}^{\text{dhn}},\ k \in \mathbb{S}_k \tag{2.26}$$

Therein, the control path edge incidence matrix elements $R_{cp,e}$ are given as:

$$R_{cp,e} = \left\{ \begin{array}{r} 1, \text{ if } e \text{ is directed as } cp, \\ -1, \text{ if } e \text{ is directed in opposition to } c,p \\ 0, \text{ if } e \text{ is not element of } cp, \end{array} \right\} \ \forall\, cp \in \mathbb{S}_{cp}^{\text{dhn}}, e \in \mathbb{S}_e^{\text{dhn}} \tag{2.27}$$

Pump The pumps are utilized in DHNs to create the necessary differential pressure to circulate the water throughout the other components of the networks. These are typically powered by electricity and local pump controllers, which adjust the motor rotation speed in order to achieve the desired differential pressure. Within the model this is represented by [OUGP16]:

$$\Delta p_{e,k} = \beta_{e,k}, \qquad \forall\, e \in \mathbb{S}_e^{\text{pump}},\ k \in \mathbb{S}_k \tag{2.28}$$

where $\mathbb{S}_e^{\text{pump}}$ describes all edges holding pumps and β can either be a fixed parameter in the case of a predefined pressure difference over the pump[38] or a variable which is obtained from the solution process of the optimization presented in Chapter 3. Note, that within this model form further forms of pump operation can be represented. Using control paths, described in the previous paragraph, a certain differential pressure over several sequential edges of the

[38] Defined by the control room of the DHN, as often performed today.

network can be achieved. Also, the pressure p at a certain node i of the network[39] or the amount of mass flow $\dot{m}$ thought the pump [Opp15, NTJK20] can be controlled, by adding additional constraints to the optimization problem, as $p_{i,k} = p_i^{\text{pre}}$ or $\dot{m}_{e,k} = \dot{m}_{\text{e}}^{\text{pre}}$ $\forall k \in \mathbb{S}_k$. Where p_i^{pre} and $\dot{m}_{\text{e}}^{\text{pre}}$ represent the predefined set values of the absolute pressure and the mass flow in either of these cases.

Differential Pressure Regulator DPRs help to improve the efficiency of DHN operation. These components represent valves, which are opened and closed in order to keep the pressure drop over a control path, e.g. a consumer sub station, constant. This provides two important functionalities. First, large pressure changes can affect the operational functionality of simple valve controls which can cause unwanted pressure oscillations. The later can cause material fatigue and inefficient valve operation [Opp15, p. 47]. Second, the utilization of DPRs helps to implement an efficient hydraulic balance within the DHN [NTJK20, pp. 146-147]. This accomplishes that the mass flows through consumers farther away from the producers are streamed through with a similar power as the consumers which are very close to the producers. The model is given as:

$$\Delta p_{e,k} = \beta_{e,k}, \qquad \forall\, e \in \mathbb{S}_{\text{e}}^{\text{dpr}},\ k \in \mathbb{S}_k \tag{2.29}$$

the value of the variable β is obtained from the solution of the optimization problem presented in Chapter 3 and $\mathbb{S}_{\text{e}}^{\text{dpr}}$ describes the set of all edges with DPRs.

Valve Together with pumps and DPRs, valves are the central active components defining the hydraulic effects with in a DHN. The pressure difference over this component type is dependent on the mass flow through the valve and can be modeled as [OUGP16]:

$$\Delta p_{e,k} = \mu_{e,k} \dot{m}_{e,k} |\dot{m}_{e,k}|_{\Delta\varepsilon}, \qquad \forall\, e \in \mathbb{S}_{\text{e}}^{\text{vlv}},\ k \in \mathbb{S}_k \tag{2.30}$$

$$\mu_{e,k} = \frac{\Delta p_0 (c^{\text{temp}})^2}{(K_{e,k})^2 \rho_0 \rho^{\text{w}}}, \qquad \forall\, e \in \mathbb{S}_{\text{e}}^{\text{vlv}},\ k \in \mathbb{S}_k \tag{2.31}$$

where the pressure coefficient μ is dependent on the valve flow factor K, its pressure reference Δp_0, the density of water ρ^{w}, the density reference ρ_0, and the temporal conversion factor $c^{\text{temp}} = 3600\,\text{s}\,\text{h}^{-1}$. Further, the set of all valves, is stated as a subset of all edges as $\mathbb{S}_{\text{e}}^{\text{vlv}}$. Note, that the operational limits of K refer to open and closed valve state, see Section 3.3.1.

Heat Exchanger of Consumer or Producer The pressure difference over heat exchangers of consumers and producers can be modeled as:

$$\Delta p_{e,k} = \mu_{e,k} \dot{m}_{e,k} |\dot{m}_{e,k}|_{\Delta\varepsilon}, \qquad \forall\, e \in \mathbb{S}_{\text{e}}^{\text{exch}},\ k \in \mathbb{S}_k \tag{2.32}$$

$$\mu_{e,k} = \frac{1}{\rho^{\text{w}} (d_e)^4} \xi_e, \qquad \forall\, e \in \mathbb{S}_{\text{e}}^{\text{exch}},\ k \in \mathbb{S}_k \tag{2.33}$$

[39] E.g. the point with lowest differential pressure between supply and return network, which often describes the consumer with the largest distance to heat production.

as given in [Köc00] for all e which are in the set of edges containing heat exchangers $\mathbb{S}_e^{\text{exch}}$. The friction factor ξ and the diameter d_e, and thereby μ, are determined by parameter identification procedures. Note, that this comprises models, as implemented in [OUGP16, Opp15], where $\mu = 0$ is used.

Pipeline For all edges e representing pipelines $\mathbb{S}_e^{\text{pipe}}$ the model is defined as:

$$\Delta p_{e,k} = \beta_{e,k} + \mu_{e,k}\dot{m}_{e,k}|\dot{m}_{e,k}|_{\Delta\varepsilon}, \qquad \forall\, e \in \mathbb{S}_e^{\text{pipe}},\ k \in \mathbb{S}_k \tag{2.34}$$

$$\beta_{e,k} = \rho^{\text{w}} g^{\text{acc}} \Delta h, \qquad \forall\, e \in \mathbb{S}_e^{\text{pipe}},\ k \in \mathbb{S}_k \tag{2.35}$$

$$\mu_{e,k} = \frac{8}{\pi^2 d_e^4 \rho^{\text{w}}}\left(\frac{L_e}{d_e}\xi_{e,k} + \psi_e\right), \qquad \forall\, e \in \mathbb{S}_e^{\text{pipe}},\ k \in \mathbb{S}_k \tag{2.36}$$

Therein, the differential pressure is dependent on the height difference Δh between both nodes at the beginning and end of the respective pipeline and the gravitational acceleration g^{acc} [Köc00, Trö99]. The pressure coefficient μ is dependent on the diameter d and the length L of the pipeline, the density of the water ρ^{w} and the friction factor ξ. Also further pressure loss coefficients, e.g. resulting from curvature are accumulated in ψ. The Darcy friction factor ξ itself is highly dependent on the flow conditions, defined by the Reynolds number Re. For laminar flows, which are found beneath the critical Reynolds numbers of $\text{Re}^{\text{crit}} = 2300$ for water, the Darcy friction factor can be calculated by [VDI13]:

$$\xi_{e,k} = \frac{64}{\text{Re}_{k,e}}, \qquad \forall\, e \in \mathbb{S}_e^{\text{pipe}},\ k \in \mathbb{S}_k \tag{2.37}$$

for larger Reynolds numbers, $\text{Re} > \text{Re}^{\text{crit}}$, which describe the transition zone towards turbulent flow and turbulent flow conditions[40], the Prandtl-Colebrook equation is utilized to determine the Darcy friction factor as [Ick95]:

$$\frac{1}{\sqrt{\xi_{e,k}}} = -2\log_{10}\left(\frac{2.51}{\text{Re}_{e,k}\sqrt{\xi_{e,k}}} + \frac{r_e}{3.7d_e}\right), \qquad \forall\, e \in \mathbb{S}_e^{\text{pipe}},\ k \in \mathbb{S}_k \tag{2.38}$$

$$\approx -1.8\log_{10}\left(\left(\frac{r_e}{3.71d_e}\right)^{1.11} + \frac{6.9}{\text{Re}_{k,e}}\right), \qquad \forall\, e \in \mathbb{S}_e^{\text{pipe}},\ k \in \mathbb{S}_k \tag{2.39}$$

$$\approx -2\log_{10}\left(\left(\frac{r_e}{3.71d_e}\right) + \frac{5.74}{(\text{Re}_{k,e})^{0.9}}\right), \qquad \forall\, e \in \mathbb{S}_e^{\text{pipe}},\ k \in \mathbb{S}_k \tag{2.40}$$

Two approximations of (2.38) are given above. The first is the Haaland and the second the Swamee approximation, which both enable to replace the implicit Prandtl-Colebrook equation by an explicit alternative [Brk11]. The roughness values of the pipeline wall r is typically around $r \approx 0.01$ mm in DHN pipelines [NTJK18, p. 118].

40 The transition zone between laminar and turbulent flow is defined around $2300 < \text{Re} < 5000$, see [Opp15, p. 33] for more details, as well as [VDI13, p. 1225] and [NTJK20, p. 118].

The Reynolds number itself is defined as [VDI13]:

$$\mathrm{Re}_{e,k} = \frac{d_e v_{e,k}}{\nu}, \qquad \forall\, e \in \mathbb{S}_e^{\mathrm{pipe}},\ k \in \mathbb{S}_k \tag{2.41}$$

$$\approx \frac{d_e |v_{e,k}|_{\Delta\varepsilon}}{\nu}, \qquad \forall\, e \in \mathbb{S}_e^{\mathrm{pipe}},\ k \in \mathbb{S}_k \tag{2.42}$$

In order to calculate the Reynolds number of a pipeline correctly, independent of the flow direction, the absolute value approximation is used in equation (2.42) as in equation (2.20) and [Trö99]. It is dependent on the flow velocity v and the kinematic viscosity of water ν. In normal DHN operation the absolute values of the flow velocities v in the pipelines range from $0.8\,\mathrm{m/s}$ to $3\,\mathrm{m/s}$ [Glü85b, p. 29], however to also cover less likely lower flows a lower limit of $1 \times 10^{-3}\,\mathrm{m/s}$ is regarded here. The kinematic viscosity of water can range from $0.2366 \times 10^{-6}\,\mathrm{m^2/s} \leq \nu \leq 1.8 \times 10^{-6}\,\mathrm{m^2/s}$ for typical pressure and temperature ranges of DHN states, see material properties of water in Section 2.3.1. Thus, possible Reynolds numbers[41] are within $2.722 \times 10^2 \leq \mathrm{Re} \leq 6.3339 \times 10^6$. This leads to:

Assumption 2.3. *The Prandtl-Colebrook equation, stated in equation* (2.38), *can be adequately represented by the Haaland or the Swamee approximation, given in* (2.39) *and* (2.40), *for all possible Reynolds numbers.*

This is due to the fact, that for $\mathrm{Re} > \mathrm{Re}^{\mathrm{crit}}$ the deviation between (2.38) and (2.39) or (2.40) is small, and the larger deviation for lower Reynolds numbers has a minor effect, as the overall pressure drop becomes negligible in this case, compared to other pipelines with higher flow rates [Trö99]. To underline these facts, the following example quantifies the impact of the utilized approximation. The calculations are performed for a pipeline with a diameter of $d = 0.5\,\mathrm{m}$, a length of $L = 100\,\mathrm{m}$, and a roughness value of $r = 0.01\,\mathrm{mm}$. Further a constant density of water of $\rho^{\mathrm{w}} = 960\,\mathrm{kg/m^3}$ is used, due to the small relative changes of the water density for DHN operation conditions, see Assumption 2.1 in Section 2.3.1. The resulting pipeline friction factors ξ are shown in Figure 2.7. Note, that the values of the friction factor resulting from the Haaland and the Swamee approximation only show very small deviations below $1.4\,\%$. Therefore these approximations are hard to distinguish before Figure 2.10.

Based on the friction factors ξ, the pressure coefficients μ can be determined, which are shown in Figure 2.8. This graph shows the desired reference as the *Laminar and Prandtl-Colebrook* case, which is based on a calculation of ξ with (2.37) for $\mathrm{Re} < \mathrm{Re}^{\mathrm{crit}}$ and (2.38) for $\mathrm{Re} > \mathrm{Re}^{\mathrm{crit}}$. The differential pressure over the entire pipeline calculated by (2.34) for different flow velocities is shown in Figure 2.9. Therefore, the differential pressure is once calculated with the desired reference values of μ, given as the laminar and Prandtl-Colebrook case in Figure 2.8[42] and once with the values of μ calculated with the Haaland approximation in (2.39). For increasing flow velocities, the lowest possible Reynolds numbers increase as well, see (2.41). These limits $\mathrm{Re}^{\mathrm{min}}$ are now defined by the maximal value of the kinematic viscosity of water $\nu^{\mathrm{max}} = 1.8 \times 10^{-6}\,\mathrm{m^2/s}$. To depict the limits of the possible differential pressures and the deviations of the later, for the different flow velocities, the aforementioned $\mathrm{Re}^{\mathrm{min}}$ limits

41 Given minimal flow velocities of $v = 1 \times 10^{-3}\,\mathrm{m/s}$.

42 Plotted in green color.

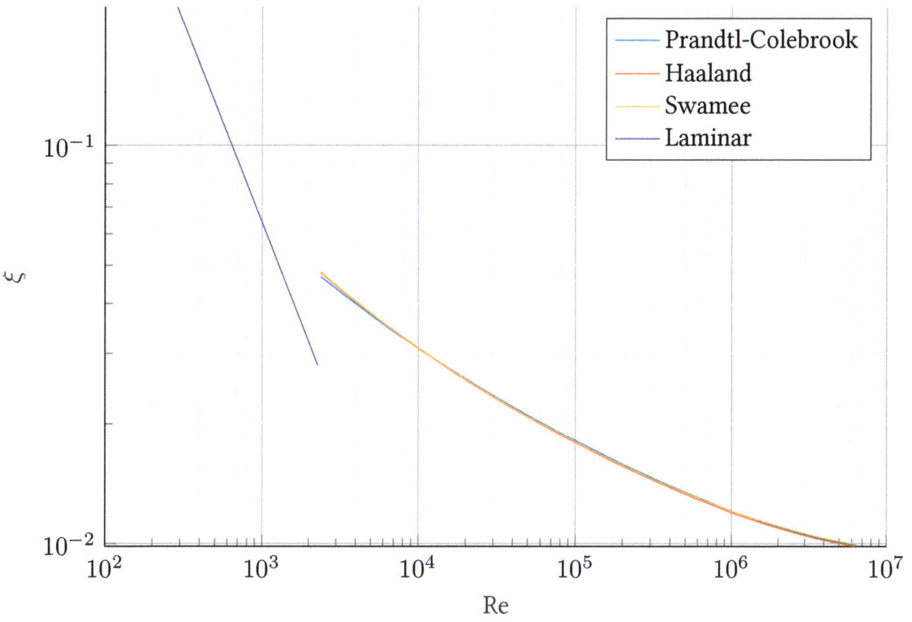

Figure 2.7: Darcy friction factor for possibly occurring Reynolds numbers in DHN operation conditions, including laminar and turbulent flows.

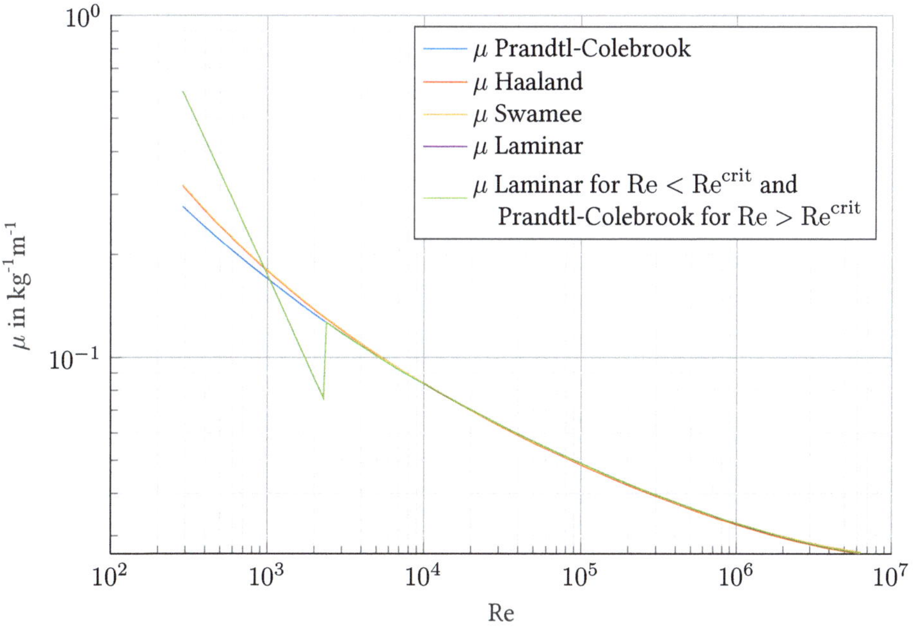

Figure 2.8: Values of the pipeline pressure coefficient μ for possibly occurring Reynolds numbers in DHN operation conditions resulting from the different Darcy friction factor calculation forms.

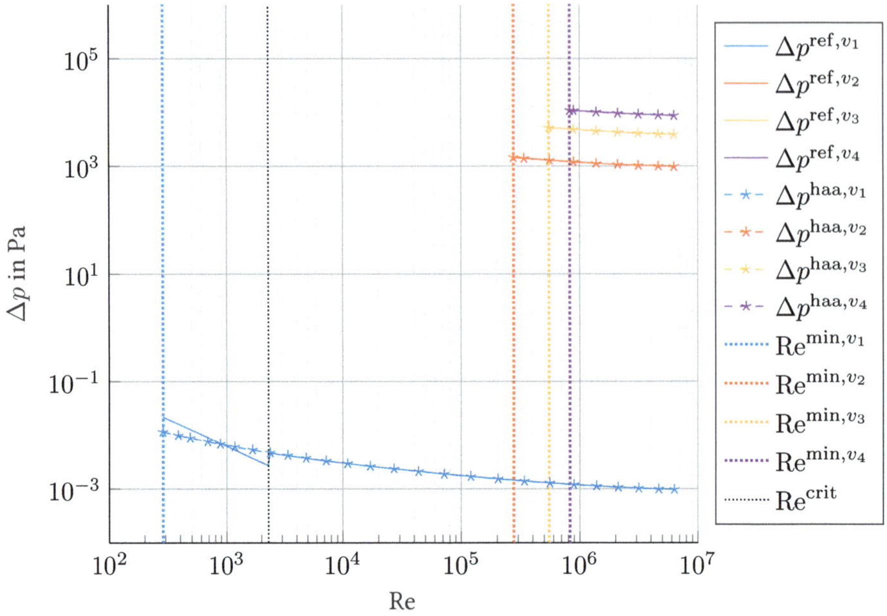

Figure 2.9: Differential pressure Δp for different flow velocities $v_1 = 1 \times 10^{-3}$ m/s, $v_2 = 1$ m/s, $v_3 = 2$ m/s and $v_4 = 3$ m/s for possibly occurring Reynolds numbers in DHN operation conditions, obtained from different calculations of the friction factor. Note, that due to the small differences between the Haaland and the Swamee approximation, only the prior is shown here. However both are further compared in the following analysis.

are shown as well in Figure 2.9 and 2.10. The resulting absolute deviation Δp^{dev} between the pressure drop resulting from the exact calculation from the pressure drop reference Δp^{ref} with the stated approximations Δp^{haa} or $\Delta p^{\mathrm{swamee}}$ is given in Figure 2.10. The relative error arising from the used approximations on the calculated differential pressure over the pipeline is given in Table 2.2 and 2.3. The quantitative study shows, that the hydraulic pipeline model using either the Swamee or the Haaland approximation for all flow conditions is very accurate, with errors below 1.3 %, for flow velocities $v > 1$ m/s. Further, for flow velocities of $v^{\mathrm{crit}} = 8.28 \times 10^{-3}$ m/s the critical Reynolds number is definitely reached. Therefore, it can be understood, that in DHNs the non laminar flow conditions represent the prevailing case. Much more, based on typical flow velocities turbulent flow can be assumed in DHN pipelines [NTJK20, p. 117]. For $\mathrm{Re} > \mathrm{Re}^{\mathrm{crit}}$ the relative error instantly decreases to the lower single-digit percent range. Thus, the used approximations lead to very exact results in the prevailing cases. In the case of very low flow velocities $v < v^{\mathrm{crit}}$ the resulting pressure drop over the pipeline becomes negligibly small, as it is a factor 10^6 smaller than the largest possible pressure drop. Therefore, the errors brought to the rest of the model, e.g. pressure drops of other pipelines, are then also neglectable.

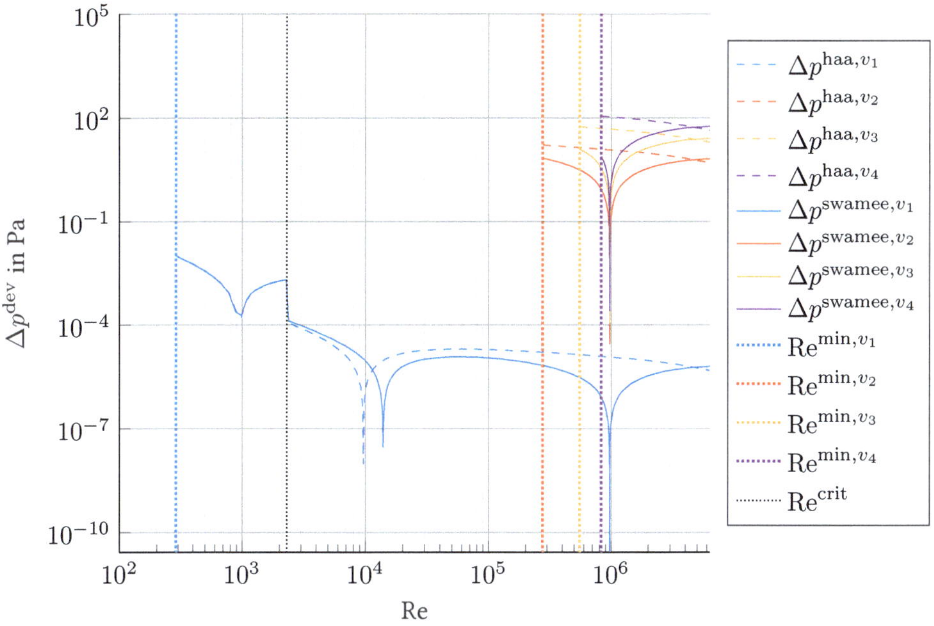

Figure 2.10: Pressure difference deviation occurring from different calculations of the friction factor for differing velocities for possibly occurring Reynolds numbers in DHN operation conditions

Table 2.2: Set of relevant values for estimation of occurring error through the used Swamee approximation for the Darcy friction factor calculation.

Flow velocity v	Maximal absolute deviation in pressure drop $\Delta p^{\text{dev}} = \lvert\Delta p^{\text{ref}} - \Delta p^{\text{swamee}}\rvert$	Pressure drop at maximal deviation Δp^{ref}	Relative error $\Delta p^{\text{dev}}/\Delta p^{\text{ref}}$
1×10^{-3} m/s	0.0101 Pa	0.0213 Pa	47.42 %
1 m/s	6.583 Pa	1431 Pa	0.46 %
2 m/s	11.8 Pa	5079 Pa	0.23 %
3 m/s	54.93 Pa	8552 Pa	0.64 %

Storage The storage model is highly dependable on the type of the storage and its form of operation. We restrict ourselves to unpressurized sensible[43] hot water storages [NTJK20], which are directly connected to the grid, as sensible hot water storages are the most used form in practice, due to the low costs and the high level of technological maturity [NTJK18] [CMT+17, p. 73f]. Hereby, there is no hydraulic separation between the grid and the storage.

43 "In sensible heat storage, a storage medium is heated or cooled without phase change." [NTJK20, p. 37]

Table 2.3: Set of relevant values for estimation of occurring error through the used Haaland approximation for the Darcy friction factor calculation.

Flow velocity v	Maximal absolute deviation in pressure drop $\Delta p^{\mathrm{dev}} = \lvert\Delta p^{\mathrm{ref}} - \Delta p^{\mathrm{haa}}\rvert$	Pressure drop at maximal deviation Δp^{ref}	Relative error $\Delta p^{\mathrm{dev}}/\Delta p^{\mathrm{ref}}$
1×10^{-3} m/s	0.0101 Pa	0.0213 Pa	47.42 %
1 m/s	15.86 Pa	1431 Pa	1.11 %
2 m/s	53.89 Pa	5079 Pa	1.06 %
3 m/s	108.7 Pa	8552 Pa	1.27 %

Therefore, the hydraulics of the storage can be modeled similarly as a pipeline with a large diameter.

$$\Delta p_{e,k} = \beta_{e,k} + \mu_{e,k} \dot{m}_{e,k} |\dot{m}_{e,k}|_{\Delta\varepsilon}, \qquad \forall\, e \in \mathbb{S}_e^{\mathrm{stor}},\ k \in \mathbb{S}_k \tag{2.43}$$

$$\beta_{e,k} = \rho^{\mathrm{w}} g \Delta h, \qquad \forall\, e \in \mathbb{S}_e^{\mathrm{stor}},\ k \in \mathbb{S}_k \tag{2.44}$$

The values of the pressure coefficient μ depend on the special design form e.g. if plates are used to help preserve the stratification. The coefficient is determined by a parameter identification procedure, with $\mu \geq 0$. Further, $\Delta p_{e,k}$ can become ≥ 0 and ≤ 0 based on the two possible cases, charging and discharging. The set of all edges containing thermal storage devices is given by $\mathbb{S}_e^{\mathrm{stor}}$.

2.3.3 Thermal model

The thermal model describes the temperature changes occurring in the different network components and the storage dynamics of pipelines and Thermal Energy Storage Systems (TESS).

Node In network graphs in general and for DHNs specifically, nodes are the components connecting multiple edges at a certain location. The thermal node models determine the node temperatures, which are important as they define the temperatures of the water entering the connected edges with water inflow. These node temperatures are calculated by using the following simplification:

Assumption 2.4. *Water flowing into a DHN node i from multiple edges e undergoes perfect instant intermixture and therefore the out flowing water has a homogeneous temperature T.*

Note, that this assumption is used in the prevailing amount of literature in the field of optimal DHN operation, see e.g. [DLS+19, LWS+15, CWW+18b, Trö99] and also in many simulation models [LWJB16, OUGP16, Ick95]. The thermal balance equation is given below for two cases. The first is the simpler case, where only edges with predefined flow directions are connected to the node. The set of nodes i for which this condition holds is given as the set of all nodes excluding the nodes where edges with possibly VMFDs are connected, thus $\mathbb{S}_i^{\mathrm{dhn}} \wedge i \notin \mathbb{S}_i^{\mathrm{vmfd}}$. This very common equation in DHN modeling can be found e.g. in [LWS+15]:

$$\sum_{e\in\mathbb{S}_e^{\mathrm{dhn}}} A_{i,e}^{\mathrm{dhn},-} \dot{m}_{e,k} T_{i,k} = \sum_{e\in\mathbb{S}_e^{\mathrm{dhn}}} (A_{i,e}^{\mathrm{dhn},+} \dot{m}_{e,k} T_{e,k}^{\mathrm{out}}), \qquad \forall\, i \in \mathbb{S}_i^{\mathrm{dhn}} \wedge i \notin \mathbb{S}_i^{\mathrm{vmfd}}, k \in \mathbb{S}_k \tag{2.45}$$

The node temperature T_i and the outlet temperatures T^{out} of the edges with water flowing into the node are found in the above equation, as well as the elements of the incoming and leaving edge node incidence matrices $\boldsymbol{A}^{\mathrm{dhn},+}$ and $\boldsymbol{A}^{\mathrm{dhn},-}$ of the DHN. These elements are defined as:

$$\begin{aligned} A_{i,e}^{\mathrm{dhn},+} &= \begin{Bmatrix} 1, \text{ if } A_{i,e}^{\mathrm{dhn}} = +1, \\ 0, \text{ else}, \end{Bmatrix} && \forall\, i \in \mathbb{S}_i^{\mathrm{dhn}}, e \in \mathbb{S}_e^{\mathrm{dhn}} \\ A_{i,e}^{\mathrm{dhn},-} &= \begin{Bmatrix} 1, \text{ if } A_{i,e}^{\mathrm{dhn}} = -1, \\ 0, \text{ else}, \end{Bmatrix} && \forall\, i \in \mathbb{S}_i^{\mathrm{dhn}}, e \in \mathbb{S}_e^{\mathrm{dhn}} \end{aligned} \tag{2.46}$$

The second case is defined for all nodes where at least on one connected edge a flow direction change can take place. These nodes i are consolidated in $\mathbb{S}_i^{\mathrm{vmfd}}$. In this case the heat balance equation considers all possible flow conditions, as given in [Trö99]:

$$\begin{aligned} &\left(\sum_{e\in\mathbb{S}_e^{\mathrm{dhn}}} A_{i,e}^{\mathrm{dhn},-} \max\left(\dot{m}_{e,k}, 0\right) + \sum_{e\in\mathbb{S}_e^{\mathrm{dhn}}} A_{i,e}^{\mathrm{dhn},+} \max\left(-\dot{m}_{e,k}, 0\right) \right) T_{i,k} \\ &= \sum_{e\in\mathbb{S}_e^{\mathrm{dhn}}} (A_{i,e}^{\mathrm{dhn},+} \max(\dot{m}_{e,k}, 0) T_{e,k}^{\mathrm{out}}) + \sum_{e\in\mathbb{S}_e^{\mathrm{dhn}}} (A_{i,e}^{\mathrm{dhn},-} \max(-\dot{m}_{e,k}, 0) T_{e,k}^{\mathrm{in}}), \\ &\forall\, i \in \mathbb{S}_i^{\mathrm{dhn}} \wedge i \in \mathbb{S}_i^{\mathrm{vmfd}}, k \in \mathbb{S}_k \end{aligned} \tag{2.47}$$

The right-hand side of equation (2.47) contains the sum of all temperatures entering this node, weighted by their mass flows $\dot{m}$. This has to equal the node temperature T_i, which is multiplied by the mass flows leaving the node. All possible cases, based on the varying mass flow directions in the pipelines, are distinguished by combining the incoming and leaving node edge incidence matrix elements $A_{i,e}^{\mathrm{dhn},+}$ and $A_{i,e}^{\mathrm{dhn},-}$, and the approximated maximum operator. The approximation is analogous to the absolute value function approximation in (2.20) and defined as:

$$\max(\dot{m}_{e,k}, 0) \approx \frac{\sqrt{\dot{m}_{e,k}^2 + \Delta\varepsilon} + \dot{m}_{e,k}}{2} \tag{2.48}$$

Input temperature of all components except nodes, storages and pipes Assuming the utilization of check valves[44] in all components except nodes, all components in storages systems[45], and pipes, the flow direction can be modeled as unidirectional through these edges.

[44] Valves, often utilized in water networks, which can only be streamed through in one direction.

[45] These include a valve, a pump and the heat exchanger of a storage system.

Thus, the input temperature of the respective components can be defined by:

$$T^{\text{in}}_{e,k} = \sum_{i \in \mathbb{S}^{\text{dhn}}_i} A^{\text{dhn},-}_{i,e} T_{i,k},$$
$$\forall\, e \in \mathbb{S}^{\text{dhn}}_e \wedge e \notin \mathbb{S}^{\text{pipe}}_e \wedge e \notin \mathbb{S}^{\text{stor}}_e \wedge e \notin \mathbb{S}^{\text{vlv,stor}}_e \wedge e \notin \mathbb{S}^{\text{pump,stor}}_e,\ k \in \mathbb{S}_k \tag{2.49}$$

where the set of all storage valves and pumps are given by $\mathbb{S}^{\text{vlv,stor}}_e$ and $\mathbb{S}^{\text{pump,stor}}_e$.

Pump The mechanical degree of efficiency of pumps η^{pump} describes the rate of coupling power to the delivery rate. Assuming all losses will heat up water, it can be used to describe the temperature change over a pump, as follows[46] [Opp15]:

$$T^{\text{out}}_{e,k} - T^{\text{in}}_{e,k} = |\Delta p_{e,k}| \frac{1 - \eta^{\text{pump}}}{\eta^{\text{pump}} \rho^{\text{w}} c^{\text{w}}}, \qquad \forall\, e \in \mathbb{S}^{\text{pump}}_e, k \in \mathbb{S}_k \tag{2.50}$$

The following example will show that it is possible to neglect the temperature gain over a pump in a DHN due to the small resulting error. For a given DHN pump [DHN] with a minimal mechanical efficiency η^{pump} of 30 %, a maximal differential pressure Δp of 16 bar, a minimal specific heat capacity of 4.182 kJ/(kg K), and a minimal density of water ρ^{w} of 939.2 kg/m^3 the resulting temperature gain over the pump is given in Figure 2.11. Therefrom, it can be seen that the maximal temperature change is below 1 K and can thus be neglected for simplicity reasons. Based on the above, the temperature gain over a pump can be neglected, and equation (2.50) can be simplified to [Trö99]:

$$T^{\text{out}}_{e,k} = T^{\text{in}}_{e,k}, \qquad \forall\, e \in \mathbb{S}^{\text{pump}}_e \wedge e \notin \mathbb{S}^{\text{pump,stor}}_e, k \in \mathbb{S}_k \tag{2.51}$$

Pumps installed to fill or unload storage systems can be operated in both flow directions. Consequently, the in- or outlet temperature of the edge is defined by:

$$\chi_{e,k,\sigma^+} T^{\text{out}}_{e,k} + \chi_{e,k,\sigma^-} T^{\text{in}}_{e,k} = \sum_{i \in \mathbb{S}^{\text{dhn}}_i} \left(A^{\text{dhn},-}_{i,e} \chi_{e,k,\sigma^+} + A^{\text{dhn},+}_{i,e} \chi_{e,k,\sigma^-} \right) T_{i,k},$$
$$\forall\, e \in \mathbb{S}^{\text{pump,stor}}_e, k \in \mathbb{S}_k \tag{2.52}$$

The variables χ_{e,k,σ^+} and χ_{e,k,σ^-} are used to approximate the utilization of boolean integer variables, becoming 0 or 1 for positive or negative flow on the respective edge. These variables are calculated by using a continuous differentiable approximation of the Heaviside step

[46] For that matter, the flow direction is assumed to align with the edge orientation.

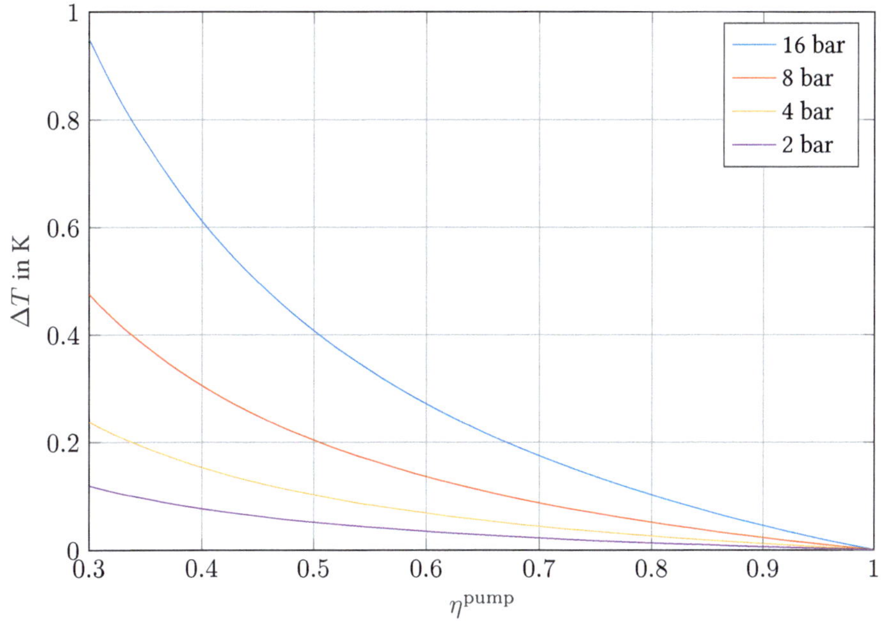

Figure 2.11: Temperature gain over DHN pump for different mechanical efficiency factors and differential pressures.

function and its at the axis of the ordinate reflected counterpart, defined as follows:

$$\begin{aligned} \chi_{e,k,\sigma^+} &= \frac{\sqrt{(\dot{m}_{e,k}+\Delta\varepsilon)^2+\Delta\varepsilon^2}-\sqrt{(\dot{m}_{e,k}-\Delta\varepsilon)^2+\Delta\varepsilon^2}+2\Delta\varepsilon}{4\Delta\varepsilon} \\ &\approx \begin{Bmatrix} 1 \text{ for } \dot{m}_{e,k}>0, \\ 0 \text{ for } \dot{m}_{e,k}<0, \end{Bmatrix} \qquad \forall\, e \in \mathbb{S}_e^{\text{stor}},\ k \in \mathbb{S}_k \qquad (2.53) \\ \chi_{e,k,\sigma^+} &= 1-\chi_{e,k,\sigma^-}, \qquad \forall\, e \in \mathbb{S}_e^{\text{stor}},\ k \in \mathbb{S}_k \qquad (2.54) \end{aligned}$$

Valve The temperature gain[46] over valves, can be calculated by [Ick95]:

$$T_{e,k}^{\text{out}} - T_{e,k}^{\text{in}} = \frac{|\Delta p_{e,k}|}{\rho^{\text{w}} c^{\text{w}}}, \qquad \forall\, e \in \mathbb{S}_e^{\text{vlv}}, k \in \mathbb{S}_k \qquad (2.55)$$

For a theoretically maximal possible pressure difference of 25 bar occurring in DHNs [ZWW$^+$21], the temperature change would be below 0.64 K[47]. Thus, the thermal models of valves can be simplified to [Trö99]:

$$T_{e,k}^{\text{out}} = T_{e,k}^{\text{in}}, \qquad \forall\, e \in \mathbb{S}_e^{\text{vlv}} \wedge e \notin \mathbb{S}_e^{\text{vlv,stor}}, k \in \mathbb{S}_k \qquad (2.56)$$

[47] The calculation is based on the same values for ρ^{w} and c^{w} as in Subsection 2.3.3.

For valves used in storage systems, the input or output temperature is defined as follows, in order to represent both flow directions in the model:

$$\begin{aligned}\chi_{e,k,\sigma^+} T^{\text{out}}_{e,k} + \chi_{e,k,\sigma^-} T^{\text{in}}_{e,k} \\ = \sum_{i \in \mathbb{S}^{\text{dhn}}_i} \left(A^{\text{dhn},-}_{i,e} \chi_{e,k,\sigma^+} + A^{\text{dhn},+}_{i,e} \chi_{e,k,\sigma^-} \right) T_{i,k}, \qquad \forall\, e \in \mathbb{S}^{\text{vlv,stor}}_e, k \in \mathbb{S}_k \end{aligned} \tag{2.57}$$

Differential Pressure Regulator DPRs are special valves which are controlled to keep the pressure difference over a control path constant. Still, their buildup does not deviate from other valves. Thus, they can be modeled as regular valves [Opp15], see equation (2.56):

$$T^{\text{out}}_{e,k} = T^{\text{in}}_{e,k}, \qquad \forall\, e \in \mathbb{S}^{\text{dpr}}_e, k \in \mathbb{S}_k \tag{2.58}$$

Consumer and Producer The heat power supply or demand Φ of producers, energy converters, and consumers is defined by the following equation, throughout the existing literature, as e.g. given in [LWJB16]:

$$\Phi_{n,k} = c^{\text{w}} \sum_{e \in \mathbb{S}^n_e} M^{\text{dhn}}_{e,n} \dot{m}_{e,k} (T^{\text{out}}_{e,k} - T^{\text{in}}_{e,k}), \qquad \forall\, n \in \mathbb{S}^{\text{exch}}_n \wedge n \notin \mathbb{S}^{\text{stor}}_n, k \in \mathbb{S}_k \tag{2.59}$$

Note, that opposed to the State of the Art (SOTA) we have included an edge NP matrix element $M^{\text{dhn}}_{e,n}$, as defined in equation (2.3), into the equation (2.59) above. Thereby, the needed sum over the set of all edges which have NPs connected $\mathbb{S}^n_e$, is used, to enable a clear mapping of an NP to the respective edge in the DHN .

Storage As discussed in Subsection 2.3.2 sensible hot water storages are regarded in this work. These TESS typically store large amounts of water, in order to store heat for days (short term TESS) up to seasons (long-term TESS) [GV19]. Short term storages for cities usually have capacities of around $15\,000\,\text{m}^3$ [GV19] up to $30\,000\,\text{m}^3$ [Kie16]. Further, they contain baffle plates which enhance the stratification process. In these storage systems, a thermocline layer separates a layer of hot water at the top of the storage from a layer of cold water at the bottom [SDPT16]. Therefore, within the application presented here, it is possible to model the temperature entering and leaving the TESS T^{in} and T^{out} by constant temperature values $T^{\text{top}}_{n,k}$ and $T^{\text{bot}}_{n,k}$ found at the top and bottom of the storage. Concerning the large amounts of water within these systems and the stratification, the values $T^{\text{top}}_{n,k}$ and $T^{\text{bot}}_{n,k}$ are assumed to remain constant over the simulation/prediction horizon[48] and are updated with measurement data after every time step. Following the aforementioned, for discharging and charging the

[48] As the presented model can be applied for simulation or optimization of CEPDHNs, simulation and prediction horizon are used both here, referring to the same set of time steps $\mathbb{S}_k$.

temperature of the water leaving the storage is defined by:

$$T^{\text{out}}_{e,k} = \sum_{n \in \mathbb{S}^{\text{dhn,stor}}_n} M^{\text{dhn}}_{e,n} T^{\text{top}}_{n,k}, \qquad \forall\, e \in \mathbb{S}^{\text{stor}}_e, k \in \mathbb{S}_k \tag{2.60}$$

$$T^{\text{in}}_{e,k} = \sum_{n \in \mathbb{S}^{\text{dhn,stor}}_n} M^{\text{dhn}}_{e,n} T^{\text{bot}}_{n,k}, \qquad \forall\, e \in \mathbb{S}^{\text{stor}}_e, k \in \mathbb{S}_k \tag{2.61}$$

The heat input or output is defined as in [NTJK18]. Note, that this equation is formulated for charging and discharging here, with the respective efficiency factors for the charging $\eta^{\text{stor,chr}}$ and the discharging $\eta^{\text{stor,dchr}}$ process, as given in [VSG+17][49]:

$$\begin{aligned}\Phi_{n,k} =& c^{\text{w}} \sum_{e \in \mathbb{S}^{\text{stor}}_e} M^{\text{dhn}}_{e,n} |\dot{m}_{e,k}|_{\Delta\varepsilon} \Bigg[\Big(\eta^{\text{stor,dchr}}_n \chi_{e,k,\sigma^+} T^{\text{out}}_{e,k} + \eta^{\text{stor,chr}}_n \chi_{e,k,\sigma^-} T^{\text{in}}_{e,k} \Big) \\ &- \sum_{i \in \mathbb{S}^{\text{dhn}}_i} \Big(\eta^{\text{stor,dchr}}_n A^{\text{dhn},-}_{i,e} \chi_{e,k,\sigma^+} + \eta^{\text{stor,chr}}_n A^{\text{dhn},+}_{i,e} \chi_{e,k,\sigma^-} \Big) T_{i,k} \Bigg], \\ &\forall\, n \in \mathbb{S}^{\text{stor}}_n, k \in \mathbb{S}_k \end{aligned} \tag{2.62}$$

Note, that the approximated binary variables, χ_{σ^+} and χ_{σ^-}, calculated during the solution process from (2.57), based on the flow direction, define if the TESS is charged or discharged. In the later case, the sign of the heat power is positive $\Phi > 0$, following the active sign convention. Further, when the storage is discharged the mass flow is positive $\dot{m} > 0$, this implies, that the edge directions of the storage components, defined in the node edge incidence matrix $\boldsymbol{A}^{\text{dhn}}$, are directed from the return network toward the supply network, see Figure 4.13. The stored quantity of heat E, the State of Charge (SOC), is defined over time as in [VSG+17, WYA+15][49]:

$$E_{n,k+1} = E_{n,k} - \Delta k \Phi_{n,k} + l_{n,k}, \qquad \forall\, n \in \mathbb{S}^{\text{stor}}_n, k < |\mathbb{S}_k| \in \mathbb{S}_k \tag{2.63}$$

$$E_{n,k_1} = E^{\text{meas}}_{n,k_1}, \qquad \forall\, n \in \mathbb{S}^{\text{stor}}_n \tag{2.64}$$

The storage losses l are calculated as defined in [VSG+17] by:

$$l_{n,k} = \Delta k A^{\text{stor}}_n R^{\text{stor}}_n (T^{\text{stor,mean}}_{n,k} - T^{\text{a}}), \qquad \forall\, n \in \mathbb{S}^{\text{stor}}_n, k < |\mathbb{S}_k| \in \mathbb{S}_k \tag{2.65}$$

with the surface of the TESS A^{stor}, the heat transfer coefficient R^{stor}, and the mean temperature of the water in the storage is given by $T^{\text{stor,mean}}$.

Pipeline The choice of an adequate pipeline model is of high relevance for operational optimization tasks of CEPDHNs. This is due to the fact, that pipelines are the most common components in DHNs and thus imprecise pipeline models can lead to significant subsequent

[49] The charging and discharging efficiencies used within equation (2.63) of this work as in [VSG+17], are included in equation (2.62).

errors[50]. At the same time the amount of optimization variables used for every pipeline should be limited in order to keep the amount of optimization variables of the entire CEPDHN optimization problem manageable. Thus, in previous work of the author [MRMH21] the most common pipeline models used within CEPDHN models, namely the Node Method (NM), approximations of the NM, the Lagrangian approach and different Finite Difference (FD) approaches are qualitatively and quantitatively compared for the application within operational optimization. The results of the quantitative comparison in [MRMH21] clearly show, that the the NM brings along the highest modeling accuracy.
The downside of the NM in its original form, as presented in [Ben91], is that the pipeline model itself is dependent on the solution of integer optimization problems which need to be solved at runtime if the model is directly used within operational optimization. This turns the NLP problem into a MINLP problem, which scales badly. To circumvent the computational burden, in [LWS+15] an iterative procedure is presented which calculates the solutions of the integer optimization problems and uses these in the dispatch problem before repeating both calculations sequentially. However, the approach presented in [LWS+15] uses an approximation of the length of stay of the water masses within a pipeline and is further limited to unidirectional mass flow conditions, which are usually not given in DHNs. A different approach using a thermal pipeline model, based on a linear interpolation of node temperatures, that can be applied in the case of VMFDs is presented in [Trö99]. Still due to convergence issues, the flow direction of every edge is fixed over the entire prediction horizon to one flow direction.
The pipeline model contributed within this work, and presented below, brings together the advantages of the aforementioned approaches [Ben91, LWS+15, Trö99] and augments these by two aspects. First, a new and more precise formula for the calculation of the length of stay of a water mass within a pipeline is used, which was first presented in the authors previous work [MEKH18]. Second, the concept of the flow direction path is introduced to limit the possible combinatorial flow direction options within a pipeline over the prediction horizon.

Besides the Assumptions 2.1 and 2.2, the pipeline model is based on the following simplifications [HAH17, DAM19, Opp15]:

Assumption 2.5. *Frictional heat is negligible, the pipeline is cylindrical, has a constant heat transmission coefficient, and the ambient temperature is constant along the length of the pipeline.*

Assumption 2.6. *Heat diffusion in the axial direction can be neglected.*

Note, that the comparison of simulated results of the author based on the NM with the measurement data from [HFT+17] showed very small Root-Mean-Square Errors (RMSEs) of 0.507 °C for multiple scenarios [MRMH21]. Therefore, the assumptions above are appropriate for operational optimization. Further, within the model below the following simplification is used:

Assumption 2.7. *Pipelines do not possess a steel core.*

[50] The trivial worst case includes multiple serial pipes all modeled with a thermal model providing a too high/low output temperature.

This simplification is no limitation to the used methodology, as the model presented below could be extended to steel cores, by using equation (6) and (7) from [MRMH21], for both flow directions. Nevertheless, future fourth and fifth generation DHNs will be operated at lower temperatures [LSG$^+$17], and thus steel cores will no more be used within the respective pipelines.

The model, presented first within this work, tracks the water masses defined by a respective quadruple (e, k, ζ, σ). These indices defining a certain water mass in a pipeline e, at time step k, at position ζ, and for the flow direction σ. The position ζ represents an integer value numbering the different water masses in ascending order from the entering end of the pipeline to the outgoing end. Note, that the entering and outgoing end are in this case dependent on the flow direction. As all parameters are given for positive and negative flow direction, the entire model is set up twice, once for every flow direction. The size of the water masses is defined by the mass flow $\dot{m}$ flowing into the pipe at the respective time step. The temperature of the water masses is set equal to the corresponding temperature of the node T_i, through which the water mass entered the pipeline. The main variables and parameters, which will be fully described in the following, are depicted in Figure 2.12. Here, for simplicity reasons in this Figure 2.12 it is assumed that only the positive flow direction σ^+ case is possible for all time steps. The mass flow values defining the length of the water masses[51] shown in Figure 2.12 are predicted mass flows $\hat{\dot{m}}$. These are obtained from the last optimization of the rolling horizon approach, see Remark 2.1. Using these predicted mass flows prevents having to solve the integer optimization problems, stated in Section 2.3.3, during runtime of the optimization, presented in Chapter 3.

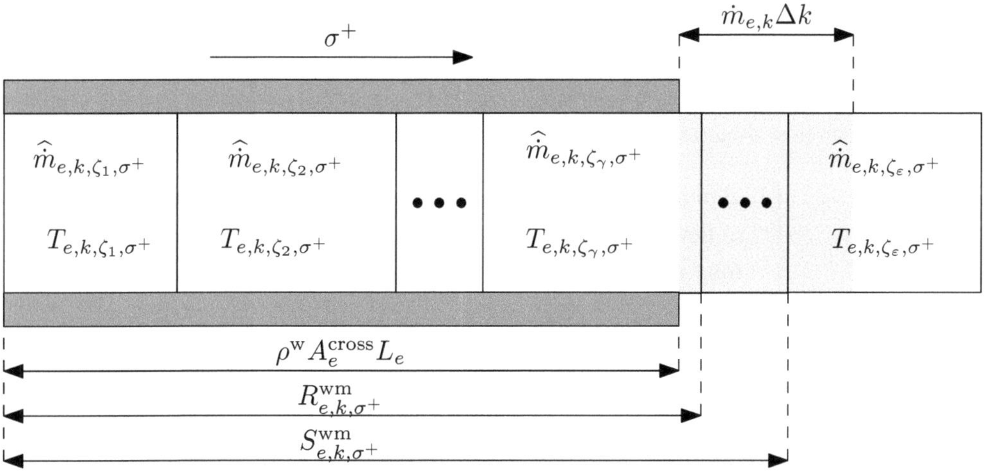

Figure 2.12: Scheme of a pipeline and relevant variables and parameters in the node method. For simplicity only the positive flow direction case σ^+ is shown here. The pipeline cross section and length are stated by A^{cross} and L. The positions ζ_γ and ζ_ε define the temporally seen last and first water masses leaving the pipeline. The water masses R^{wm} and S^{wm} describe aggregated water masses of all water elements on the positions from the beginning of the pipeline up to the one determined by γ and $\varepsilon - 1$. Note, that the calculation of these parameters is provided in Section 2.3.3.

[51] Resulting from the flow velocity during the respective time steps.

The assignment of a node temperature T_i to the respective water mass temperature $T_{e,k,\zeta,\sigma}$ is provided by the following equation:

$$T_{e,k,\zeta,\sigma} = \sum_{\tilde{\sigma}\in\mathbb{S}_\sigma} \sum_{i\in\mathbb{S}_i^{\text{dhn}}} \left(A_{i,e,\tilde{\sigma}}^{\text{dhn,sgn}} \left(\sum_{\tilde{k}\in\mathbb{S}_k} Z_{e,k,\zeta,\sigma,\tilde{k},\tilde{\sigma}} T_{i,\tilde{k}} + \sum_{k^{\text{p}}\in\mathbb{S}_{k^{\text{p}}}} Z^{\text{p}}_{e,k,\zeta,\sigma,k^{\text{p}},\tilde{\sigma}} T^{\text{p}}_{i,k^{\text{p}}} \right) \right),$$
$$\forall\, e \in \mathbb{S}_e^{\text{pipe}}, k \in \mathbb{S}_k, \zeta \in \mathbb{S}_\zeta, \sigma \in \mathbb{S}_\sigma \tag{2.66}$$

in which the set of all positions ζ and flow directions σ are stated as $\mathbb{S}_\zeta$ and $\mathbb{S}_\sigma$. The node temperatures within the current prediction horizon $T_{i,k}$ and the past time steps $T^{\text{p}}_{i,k^{\text{p}}}$ are used[52]. Further the correct mapping of the respective (past) node temperature to the distinct water mass is accomplished by three sums iterating over all flow directions σ, all nodes i and all (past) time steps $(k^{\text{p}}/)k$, and two matrices. The first is the flow direction dependent incoming/leaving node edge incidence matrix $\boldsymbol{A}^{\text{dhn,sgn}}$. This is used to select the appropriate nodes from which the water mass has entered the pipeline which is assigned by the flow direction and the regarded edge, and is defined as given below:

$$A_{i,e,\tilde{\sigma}}^{\text{dhn,sgn}} = \begin{cases} 1, & \text{if } \tilde{\sigma} = 1 \text{ and } A_{i,e}^{\text{dhn},-} = 1, \\ & \text{or if } \tilde{\sigma} = 2 \text{ and } A_{i,e}^{\text{dhn},+} = 1, \\ 0, & \text{else,} \end{cases}$$
$$\forall\, i \in \mathbb{S}_i^{\text{dhn}}, e \in \mathbb{S}_e^{\text{pipe}}, \tilde{\sigma} \in \mathbb{S}_\sigma \tag{2.67}$$

The second matrix utilized in equation (2.66) is the six dimensional *entering time step matrix* $\boldsymbol{Z}$, which is defined as $\boldsymbol{Z}^{\text{p}}$ for past time steps k^{p}. For a given quadruple (e, k, ζ, σ) defining a certain water mass[53], the entering time step matrix Z/past time step matrix $\boldsymbol{Z}^{\text{p}}$ provides the information at which time step $\tilde{k}$/past time step k^{p} this water mass has entered the pipeline and with which flow direction $\tilde{\sigma}$. The single matrix elements are given as:

$$Z_{e,k,\zeta,\sigma,\tilde{k},\tilde{\sigma}} = \begin{cases} 1, & \text{if in pipeline } e \text{, at time step } k \text{, the water mass at position } \zeta \\ & \text{of flow direction } \sigma \text{ has entered the pipeline at} \\ & \text{time step } \tilde{k} \text{ with flow direction } \tilde{\sigma}, \\ 0, & \text{else,} \end{cases}$$
$$\forall\, e \in \mathbb{S}_e^{\text{pipe}},\quad k, \tilde{k} \in \mathbb{S}_k,\quad \zeta \in \mathbb{S}_\zeta,\quad \sigma, \tilde{\sigma} \in \mathbb{S}_\sigma \tag{2.68}$$

$$Z^{\text{p}}_{e,k,\zeta,\sigma,k^{\text{p}},\tilde{\sigma}} = \begin{cases} 1, & \text{if in pipeline } e \text{, at time step } k \text{, the water mass at position } \zeta \\ & \text{of flow direction } \sigma \text{ has entered the pipeline at the past} \\ & \text{time step } k^{\text{p}} \text{ with the flow direction } \tilde{\sigma}, \\ 0, & \text{else,} \end{cases}$$
$$\forall\, e \in \mathbb{S}_e^{\text{pipe}},\quad k \in \mathbb{S}_k,\quad \zeta \in \mathbb{S}_\zeta,\quad k^{\text{p}} \in \mathbb{S}_{k^{\text{p}}},\quad \sigma, \tilde{\sigma} \in \mathbb{S}_\sigma \tag{2.69}$$

[52] Note that $\tilde{k}$ and k, as well as $\tilde{\sigma}$ and σ are defined for the same set $\mathbb{S}_k$ and $\mathbb{S}_\sigma$ respectively.

[53] Within the model, in a pipeline e, at time step k, at position ζ, and for the flow direction σ.

For a more intuitive understanding, exemplary values for entries in the matrices $\boldsymbol{Z}$ and $\boldsymbol{Z}^{\mathrm{p}}$ are provided in Example 2.1 in (2.86) and (2.87).

The lossless outlet temperature of a pipeline is calculated based on a weighted sum of the temperatures of the water masses leaving the pipeline in the current time step, see also Figure 2.12, as given in [Ben91][54]:

$$\begin{aligned} T^{\mathrm{out1}}_{e,k,\sigma} &= w_{e,k,\gamma,\sigma} T_{e,k,\gamma,\sigma} + \sum_{\zeta=\gamma+1}^{\varepsilon-1} w_{e,k,\zeta,\sigma} T_{e,k,\zeta,\sigma} + w_{e,k,\varepsilon,\sigma} T_{e,k,\varepsilon,\sigma} \\ &= \sum_{\zeta=\gamma}^{\varepsilon} w_{e,k,\zeta,\sigma} T_{e,k,\zeta,\sigma}, \qquad \forall\, e \in \mathbb{S}^{\mathrm{pipe}}_{\mathrm{e}}, k \in \mathbb{S}_k, \sigma \in \mathbb{S}_\sigma \end{aligned} \tag{2.70}$$

Note, that for both flow directions an output temperature is provided, as the index σ can stand for positive or negative flow direction σ^+ or σ^-, also see Figure 2.13. The exact calculation of the weights $w \in [0,1]$ used in the above equation (2.70) are given later in equations (2.100) to (2.102). The temperature drop occurring to water masses leaving the pipeline is calculated by equation (2.71). Heat losses rise with increasing length of stay t and difference between the lossless outlet temperature T^{out1} and the ambient temperature T^{a}. Thus, the lossy outlet temperature T^{out2} is given by [Ben91]:

$$T^{\mathrm{out2}}_{e,k,\sigma} = T^{\mathrm{a}} + (T^{\mathrm{out1}}_{e,k,\sigma} - T^{\mathrm{a}}) \exp\left(-\frac{t_{e,k,\sigma}}{\tau_e}\right), \qquad \forall\, e \in \mathbb{S}^{\mathrm{pipe}}_{\mathrm{e}}, k \in \mathbb{S}_k, \sigma \in \mathbb{S}_\sigma \tag{2.71}$$

The used loss coefficient τ_e is dependent on the cross section of the pipeline A^{cross} and its thermal resistance R^{therm} and is calculated by: $\tau_e = \rho^{\mathrm{w}} c^{\mathrm{w}} A^{\mathrm{cross}}_{\mathrm{e}} R^{\mathrm{therm}}_{\mathrm{e}}$. The temperature of the water mass leaving the pipeline in case of positive or negative flow direction σ^+ or σ^- is then matched with the leaving or entering edge temperature $T^{\mathrm{out}}_{\mathrm{e}}$ or $T^{\mathrm{in}}_{\mathrm{e}}$ by:

$$T^{\mathrm{out}}_{e,k} = T^{\mathrm{out2}}_{e,k,\sigma^+}, \qquad \forall\, e \in \mathbb{S}^{\mathrm{pipe}}_{\mathrm{e}}, k \in \mathbb{S}_k \tag{2.72}$$

$$T^{\mathrm{in}}_{e,k} = T^{\mathrm{out2}}_{e,k,\sigma^-}, \qquad \forall\, e \in \mathbb{S}^{\mathrm{pipe}}_{\mathrm{e}}, k \in \mathbb{S}_k \tag{2.73}$$

Note, that in the thermal node equations (2.45) and (2.47) only $T^{\mathrm{out}}_{\mathrm{e}}$ or $T^{\mathrm{in}}_{\mathrm{e}}$ has a relevant impact on the temperature T_i of one of the nodes connected to the pipeline i. This is due to the fact that either $\max(\dot{m}, 0)$ or $\max(-\dot{m}, 0)$ in (2.47) will become nearly zero depending on the flow direction or that only $T^{\mathrm{out}}_{\mathrm{e}}$ is taken into account in (2.45). The other temperature $T^{\mathrm{out}}_{\mathrm{e}}$ or $T^{\mathrm{in}}_{\mathrm{e}}$ respectively does not further impact the model.

After the standard rolling horizon implementation, see Remark 2.1 or Section 3.3, which is used for most of the variables in the approach presented, the rolling horizon implementation for the following parameters $Z_{e,k,\zeta,\sigma,\tilde{k},\tilde{\sigma}}$, $Z^{\mathrm{P}}_{e,k,\zeta,\sigma,k^{\mathrm{P}},\tilde{\sigma}}$, $\widehat{m}_{e,k,\zeta,\sigma}$, and $\kappa_{e,k,\zeta,\sigma}$ needs special treatment in order to adequately represent possible flow direction changes. The values in $\kappa_{e,k,\zeta,\sigma}$ represent the time steps a water mass has been in the pipeline since entering it, which are important values to determine the temperature losses occurring in the pipe. As soon as these parameters

[54] Note that in [Ben91] this equation is not given for both flow directions, also the notation is slightly different as the weights w are replaced by their exact definition, which is given later in equations (2.100) to (2.102).

are determined, the parameters dependent on $\widehat{m}_{e,k,\zeta,\sigma}$ and $\kappa_{e,k,\zeta,\sigma}$ can be recalculated for the upcoming time step. These dependent parameters[55] are needed to calculate the lossless outlet temperature of a pipeline T^{out1} with (2.70) as well as the lossy outlet temperature T^{out2} in (2.71), and will be introduced in Section 2.3.3. Before the special form of rolling horizon parameter assignment, which was developed for this model, is presented, the flow direction path concept is introduced, as it plays an important role thereby.

The flow direction path concept As stated above the temperatures and mass flows of the water masses in the pipelines are taken into account throughout the model for positive and negative flow directions. However, representing all $2^{\text{n}^{\text{ck}}}$ possible flow combinations over the entire prediction/simulation horizon[48], with the cardinality of the time steps on the prediction horizon $\text{n}^{\text{ck}} = |\mathbb{S}_k|$, would result in a large number of parameters and variables, which would have to be taken into account. Thus, only the most likely flow conditions are utilized as the

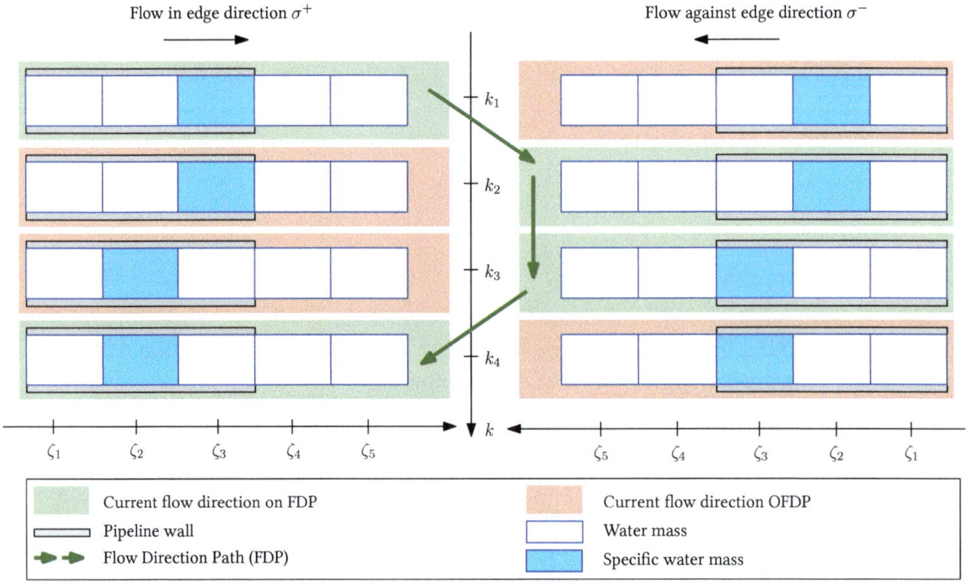

Figure 2.13: The Flow Direction Path (FDP) for a pipeline e depicted as a combination of flow directions σ over the time steps k of the prediction horizon assuming constant mass flow and thus equally sized water masses. The Opposed to Flow Direction Path (OFDP) flow directions are also shown, which are directly determined as the opposed flow directions to the FDP, for every time step k. The resulting position ζ of a specific water mass is highlighted to visualize the position changes based on the shown flow directions.

base case, while the model is also equipped with the ability to factor unexpected flow direction changes, e.g. resulting from inaccurate mass flow forecasts $\widehat{m}$.

[55] $\gamma_{e,k,\sigma}$, $\varepsilon_{e,k,\sigma}$, $R^{\text{wm}}_{e,k,\sigma}$, $S^{\text{wm}}_{e,k,\sigma}$, $w_{e,k,\zeta,\sigma}$, and $t_{e,k,\sigma}$.

Definition 2.1

*The **Flow Direction Path (FDP)** represents the predicted combination of positive and negative flow directions σ for a specific pipeline e for every time step k on the prediction horizon. These most likely flow conditions are obtained from the network flows of the last optimization. An example of the FDP is provided in Figure 2.13. The mathematical definition is given as:*

$$\sigma_{e,k}^{\text{fdp}} = sgn(\dot{m}_{e,k}), \quad \forall\, e \in \mathbb{S}_e^{\text{pipe}}, \quad k \in \mathbb{S}_k \tag{2.74}$$

*The combination of flow directions σ for the respective pipe e over the prediction horizon going in the opposite direction are defined as the **Opposed to the Flow Direction Path (OFDP)** case, which is analogously provided by:*

$$\sigma_{e,k}^{\text{ofdp}} = -sgn(\dot{m}_{e,k}), \quad \forall\, e \in \mathbb{S}_e^{\text{pipe}}, \quad k \in \mathbb{S}_k \tag{2.75}$$

Note, that the values $\dot{m}_{e,k}$ in (2.74) and (2.75) represent the initial values of the upcoming optimization. As the standard rolling horizon time shift is already performed for all variables at this point within the overall procedure, as explained in detail in Section 3.3.1.

Based on Definition 2.1 it can be understood, that the FDP and the case OFDP, as shown in Figure 2.13, are determined through the solution of the last optimization/simulation, within the rolling horizon procedure. Thus, both FDP and the OFDP case are fixed during every optimization, always taking into account the latest predictions of the last optimizations and further information as will be explained in detail in Section 3.3.1. During the upcoming solution process of the next optimization/simulation, the sign of the mass flow $\text{sgn}(\dot{m}_{e,k})$ of a certain pipeline e and distinct time step k will affect, if the FDP or the OFDP lossy outlet temperature will have a relevant impact on the node bordering the left or right side of the pipeline. The two possible cases can be seen in Figure 2.13. Regarding a certain pipe e and k_1, the first case occurs for $\text{sgn}(\dot{m}_{e,k}) = 1$ as $\sigma^{\text{fdp}} = \sigma^{+}$ the lossy outlet temperature at the right end of the pipeline $T_{e,k,\sigma^+}^{\text{out2}}$ will impact the node temperature at the right end through (2.72) which links $T_{e,k,\sigma^+}^{\text{out2}}$ to the thermal node equation (2.45). Analogously, for k_1 and $\sigma^{\text{ofdp}} = \sigma^{-}$ the left side outlet temperature $T_{e,k,\sigma^-}^{\text{out2}}$ will impact the node temperature at the left side of the pipeline. The lossy outlet temperatures are thereby dependent on the pipeline parameters stated in this section and the node temperatures, representing variables during the solution process. Changing mass flow values during the solution procedure affect which outlet temperature of both pipeline ends is used, but not which value this outlet temperature will have. This is comprehensible from (2.70), wherein the weights w are parameters, and (2.71). Therefrom it can be understood, that as soon as an unforeseen flow direction changes occur, the current and all further time steps on the prediction horizon will show deviations. As quality of the predictions are assumed to be of high quality in general and increase for time steps closer to the current point in time, usual temperature deviations are expected to be within the lower single digit range, see Section 2.99. The major advantage of this approach is that it reduces the amount of possible flow conditions taken into account from $2^{\text{n}^{\text{ck}}}$ to 2n^{ck}, which has an increasing effect for longer prediction horizons with n^{ck} time steps k. Thus, the advantages coming from the gain in computation time prevail in the context of the application.

As soon as the entire CEPDHN model is solved at a given time step k, the pipeline parameters are updated to optimally suit the latest mass flow values $\dot{m}$. Thereby, first, all pipeline parameters on the FDP are updated for all time steps of the prediction horizon of all pipelines, as stated in Case A below. As soon as these values are calculated, the parameters for the flow directions OFDP are also updated, see the description of Case B in the subsequent paragraph. These OFDP values are of relevance to the model in order to approximate the outlet temperature T^{out2} in the case of an unpredicted mass flow change.

Case A: Rolling horizon time step shift of the pipeline parameters on the flow direction path

The rolling horizon time step shift is performed in ascending order for k starting with k_1. There are two main cases that need to be differentiated between within Case A:

A.1 The flow direction on the flow direction path σ^{fdp} within a pipeline e stays the same between two consecutive time steps k. Visually speaking, as all water masses are moved ahead one position from one time step to the next, this case is referred to as "move ahead" in the following.

A.2 The flow direction on the flow direction path σ^{fdp} within a pipeline e changes between two consecutive time steps k. This case will be named "insert and copy previous time step" as parameters are assigned by utilizing measurement data or results of the last optimization for the values of the first position ζ_1[56] (*insert*) and the parameters derived for the last time step for all positions $\zeta > 1$[56] (*copy previous time step*).

These two cases are illustrated in Figure 2.14 and 2.15. The formulas are given below after two further remarks.

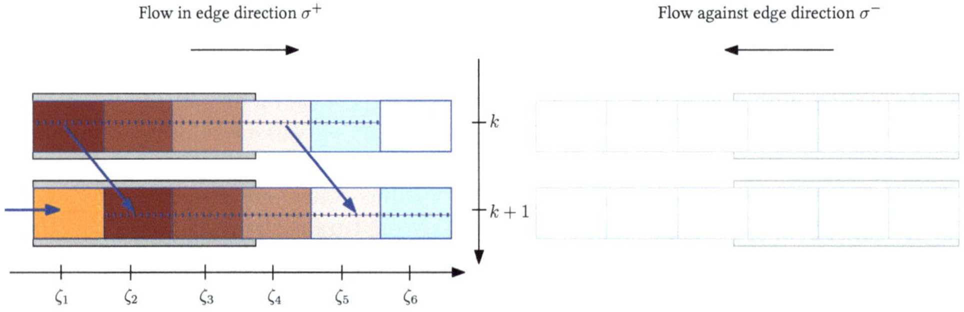

Figure 2.14: Case A.1: Basic idea of the approach if the flow direction on the flow direction path σ^{fdp} within a pipeline e stays the same between two consecutive time steps k and $k+1$, assuming constant mass flows.

56 Of the new time step.

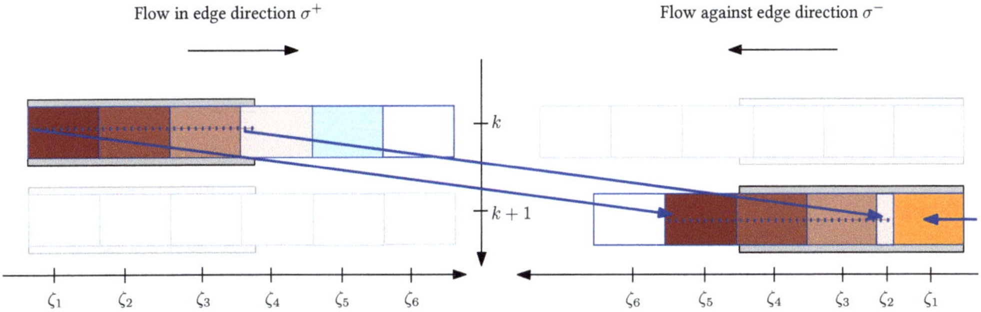

Figure 2.15: Case A.2: Basic idea of approach if the flow direction on the flow direction path σ^{fdp} within a pipeline e changes between two consecutive time steps k and $k+1$.

Note that in the following equations, multiple indices again have sub-indices. Thus, for brevity and simplicity, the following notation is defined:

Definition 2.2

If not explicitly stated differently, all sub-indices result from the so called "main indices", which are the edge e (only pipelines here), time step k, and flow direction σ. Exemplary the following expression can be simplified as[a,b]*:*

$$\widehat{\dot{m}}_{e,k-1,\zeta_{\gamma_{e,k-1,\sigma^{\mathrm{fdp}}_{e,k-1}}},\sigma^{\mathrm{fdp}}_{e,k-1}} = \widehat{\dot{m}}_{e,k-1,\zeta_{\gamma},\sigma^{\mathrm{fdp}}} \tag{2.76}$$

[a] Note, that the penultimate and last index, $\tilde{k}$ and $\tilde{\sigma}$, of the matrices $\boldsymbol{Z}$ and $\boldsymbol{Z}^{\mathrm{p}}$ do not possess sub-indices.

[b] The parameter γ in this equation describes the position of the (temporally speaking) last water mass, leaving the pipeline, see Figure 2.12.

The subsequent differing time step types are defined for clarity reasons in the following as:

Definition 2.3

The time step types occurring are:

1. *The time step on the current prediction horizon k. Note, that the time steps k are those referred to in all other equations and assignments in this work if not differently mentioned.*
2. *The (past time step/) time step at which the respective water mass entered the pipeline (k^{P}/) $\tilde{k}$.*
3. *The time steps of the previous optimization, before the currently needed rolling horizon time step shift. Thus, for clarity reasons, all parameters referring to values previous to the moving horizon time step shift are labeled with Before Rolling Horizon (BRH).*

An illustration of these definitions is provided in Figure 2.16.

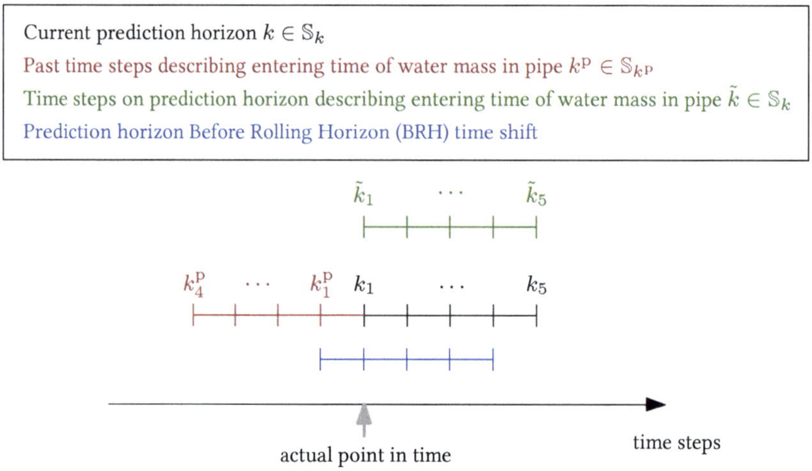

Figure 2.16: Basic example showing the relation of the different time step definitions.

The rolling horizon definition for the pipeline parameters in **Case A.1**, also referred to as *move ahead* are assigned as given below. The predicted water masses $\widehat{\dot{m}}$ are updated with last measurements or latest predictions resulting from the last optimization $\dot{m}$. Note, that only the absolute values are necessary here, as the flow direction information is already available in σ^{fdp}. For the last time step of the prediction horizon no predicted value is available at this point in time, thus this value is assumed to be identical to the mass flow of the next to last time step $k = |\mathbb{S}_k|$. The exact assignment is given by:

$$\widehat{m}_{e,k,\zeta,\sigma^{\text{fdp}}} = \begin{cases} |\dot{m}_{e,k}| \text{ for } \zeta = 1 \text{ and } k < |\mathbb{S}_k|, \\ |\dot{m}_{e,k-1}| \text{ for } \zeta = 1 \text{ and } k = |\mathbb{S}_k|, \end{cases}$$
$$\forall\, e \in \mathbb{S}_e^{\text{pipe}},\;\; k \in \mathbb{S}_k,\;\; \sigma^{\text{fdp}} \in \mathbb{S}_\sigma \tag{2.77}$$

As soon as the values of the first positions ζ_1 in the pipeline are updated, the values for the rest of the positions can be shifted ahead as expected on a moving horizon approach, as always with ascending order of k by:

$$\widehat{m}_{e,k,\zeta,\sigma^{\text{fdp}}} = \begin{cases} \widehat{m}^{\text{brh}}_{e,k_1,\zeta-1,\sigma^{\text{fdp}}} \text{ for } k = 1, \text{ and } \zeta < |\mathbb{S}_\zeta|, \\ \widehat{m}_{e,k-1,\zeta-1,\sigma^{\text{fdp}}} \text{ for } k > 1, \text{ and } \zeta < |\mathbb{S}_\zeta|, \\ \frac{1}{\Delta k}\rho^{\text{w}} A_e^{\text{cross}} L_e \;\text{ for } \zeta = |\mathbb{S}_\zeta|, \end{cases}$$
$$\forall\, e \in \mathbb{S}_e^{\text{pipe}},\;\; k \in \mathbb{S}_k,\;\; 1 < \zeta \leq |\mathbb{S}_\zeta|,\;\; \sigma^{\text{fdp}} \in \mathbb{S}_\sigma \tag{2.78}$$

The only exception to this procedure is the safety definition utilized for the last position, in order to avoid not fully filled pipelines in the model due to a limited amount of positions in the case of low mass flow rates in the past time steps. Such low mass flow rates occurring in multiple consecutive time steps are unlikely as DHN operators try to keep the water mass circulating in order to prevent accumulation of deposits, as described in last two subsections of Section 3.3.1. However, with the following definition, which artificially extends the water mass of the last position in the pipe model to the size of the entire pipeline, the model can also handle the described scenario[57].

The new entering time step matrix $\boldsymbol{Z}$ is obtained from:

$$Z_{e,k,\zeta,\sigma^{\text{fdp}},\tilde{k},\tilde{\sigma}} = \begin{cases} 1, \text{ if } \zeta > 1, k > 1, \text{ and } Z_{e,k-1,\zeta-1,\sigma^{\text{fdp}},\tilde{k},\tilde{\sigma}} = 1, \\ \quad \text{or if, } \zeta = 1,\; \tilde{k} = k \text{ and } \sigma^{\text{fdp}} = \tilde{\sigma}, \\ 0, \text{ else}, \end{cases}$$
$$\forall\, e \in \mathbb{S}_e^{\text{pipe}},\;\; k, \tilde{k} \in \mathbb{S}_k,\;\; \zeta \in \mathbb{S}_\zeta,\;\; \sigma^{\text{fdp}}, \tilde{\sigma} \in \mathbb{S}_\sigma \tag{2.79}$$

Therein, the matrix element becomes nonzero if one of the following conditions holds. First, a water mass which has entered the pipeline before, during a time step on the prediction horizon, is “pushed” one position ahead by the entering water mass. Second, a water mass is entering the pipeline at the respective time step $\tilde{k} = k$, and thus filling the first position $\zeta = 1$.

57 As long as $|\mathbb{S}_\zeta| + 1 \geq |\mathbb{S}_k|$, the calculation of γ, ε, R, and S, which are defined towards the end of the current Section, will always be possible with this definition as well.

Using this notation, the rolling horizon time step shift of the entering past time step matrix $\boldsymbol{Z}^{\mathrm{p}}$ is given by:

$$Z^{\mathrm{p}}_{e,k,\zeta,\sigma^{\mathrm{fdp}},k^{\mathrm{p}},\tilde{\sigma}} = \begin{cases} 1, \text{ if } k > 1, \text{ and } Z^{\mathrm{p}}_{e,k-1,\zeta-1,\sigma^{\mathrm{fdp}},k^{\mathrm{p}},\tilde{\sigma}} = 1, \\ \quad \text{or if } \zeta = 2,\ k = k^{\mathrm{p}} = 1, \text{and } \sigma^{\mathrm{fdp}} = \tilde{\sigma}, \\ \quad \text{or if } \zeta > 2,\ k = 1,\ k^{\mathrm{p}} > 1, \text{ and } Z^{\mathrm{p,brh}}_{e,k_1,\zeta-1,\sigma^{\mathrm{fdp}},k^{\mathrm{p}}-1,\tilde{\sigma}} = 1, \\ 0, \text{ else}, \end{cases}$$

$$\forall\, e \in \mathbb{S}^{\mathrm{pipe}}_{\mathrm{e}},\quad k \in \mathbb{S}_k,\quad k^{\mathrm{p}} \in \mathbb{S}_{k^{\mathrm{p}}},\quad \zeta \in \mathbb{S}_\zeta,\quad \sigma^{\mathrm{fdp}}, \tilde{\sigma} \in \mathbb{S}_\sigma \tag{2.80}$$

Elements of this sparse matrix are nonzero in the following three conditions:

- The first describes the intuitive realization of the moving horizon for a water mass element that was already in the pipeline in the last time step $k-1$, which has entered at a past time step k^{p}. This element is shifted one position ahead.
- The second describes the water mass which is at the second position ζ_2 of the regarded pipeline for the first time step k_1. This water mass has entered the pipeline at the last time step k^{p}_1, as the flow direction has not changed between k_1 and k^{p}_1 [58].
- The third condition is used to maintain the information about water masses which have entered in the past time steps k^{p} [59]. As the approach updates the matrix elements in ascending order of the time steps k, without this step, the propagation of the information achieved through the first condition could not be accomplished.

The definition of the amounts of time steps a respective water mass was in the pipeline can be calculated by increasing the value of the last time step by one. Only the water mass in the first position is regarded as just entering the pipeline and thus the current time step is not taken into account here. The resulting equation is given by:

$$\kappa_{e,k,\zeta,\sigma^{\mathrm{fdp}}} = \begin{cases} 0, \text{ if } \zeta = 1, \\ \kappa^{\mathrm{brh}}_{e,k_1,\zeta-1,\sigma^{\mathrm{fdp}}} + 1, \text{ for } \zeta \neq 1, \text{ and } k = 1, \\ \kappa_{e,k-1,\zeta-1,\sigma^{\mathrm{fdp}}} + 1, \text{ for } \zeta \neq 1, \text{ and } k > 1, \end{cases}$$

$$\forall\, e \in \mathbb{S}^{\mathrm{pipe}}_{\mathrm{e}},\quad k \in \mathbb{S}_k,\quad \zeta \in \mathbb{S}_\zeta,\quad \tilde{\sigma} \in \mathbb{S}_\sigma \tag{2.81}$$

In **Case A.2**, also named *insert and copy previous time step*, when the flow direction on the flow direction path σ^{fdp} changes between two consecutive time steps, the pipeline storage parameters are assigned as follows: For the first position ζ_1, measurement data or results of the last optimization are utilized for the parameter definition. Visually speaking these values are newly *inserted* into the model, see the water mass marked in orange in Figure 2.15. Additionally, the values for all positions $\zeta > 1$ are assigned from the parameters already derived for the last time step (*copy previous time step*).

58 Note that this condition can also be understood as the transition of a nonzero matrix element (representing a water mass) from $\boldsymbol{Z}$ to $\boldsymbol{Z}^{\mathrm{p}}$.

59 Note, that as indicated by Definition 2.3 the index k_1 in $\boldsymbol{Z}^{\mathrm{p,brh}}$ refers to a different time step than k_1 in $\boldsymbol{Z}^{\mathrm{p}}$. Precisely, these two indexes k_1 describe two subsequent time steps, as can be understood from Figure 2.16.

The parameters of the predicted water masses $\widehat{\dot{m}}$ are updated as follows for the first time step k_1:

$$\widehat{\dot{m}}_{e,k_1,\zeta,\sigma^{\text{fdp}}} = \begin{cases} |\dot{m}_{e,k}| & \text{for } \zeta = 1 \\ \widehat{\dot{m}}^{\text{brh}}_{e,k_1,\zeta_\gamma,\sigma^{\text{fdp}}} - \frac{1}{\Delta k}(R^{\text{wm,brh}}_{e,k_1,\sigma^{\text{fdp}}} - \rho^{\text{w}} A^{\text{cross}}_e L_e) & \text{for } \zeta = 2 \\ \widehat{\dot{m}}^{\text{brh}}_{e,k_1,\zeta_{\gamma-1},\sigma^{\text{fdp}}} & \text{for } \zeta = 3 \\ \vdots & \vdots \\ \widehat{\dot{m}}^{\text{brh}}_{e,k_1,\zeta_1,\sigma^{\text{fdp}}} & \text{for } \zeta = 2 + \gamma^{\text{brh}}_{e,k_1,\sigma^{\text{fdp}}} - 1 \\ \frac{1}{\Delta k}\rho^{\text{w}} A^{\text{cross}}_e L_e & \\ \vdots & \vdots \\ \frac{1}{\Delta k}\rho^{\text{w}} A^{\text{cross}}_e L_e & \text{for } \zeta = |\mathbb{S}_\zeta| \end{cases}$$

$$\forall\, e \in \mathbb{S}^{\text{pipe}}_e,\ \ \zeta \in \mathbb{S}_\zeta,\ \ \sigma^{\text{fdp}} \in \mathbb{S}_\sigma \tag{2.82}$$

and as follows for all further time steps:

$$\widehat{\dot{m}}_{e,k,\zeta,\sigma^{\text{fdp}}} = \begin{cases} |\dot{m}_{e,k}| & \text{for } \zeta = 1 \\ \widehat{\dot{m}}_{e,k-1,\zeta_\gamma,\sigma^{\text{fdp}}} - \frac{1}{\Delta k}(R^{\text{wm}}_{e,k-1,\sigma^{\text{fdp}}} - \rho^{\text{w}} A^{\text{cross}}_e L_e) & \text{for } \zeta = 2 \\ \widehat{\dot{m}}_{e,k-1,\zeta_{\gamma-1},\sigma^{\text{fdp}}} & \text{for } \zeta = 3 \\ \vdots & \vdots \\ \widehat{\dot{m}}_{e,k-1,\zeta_1,\sigma^{\text{fdp}}} & \text{for } \zeta = 2 + \gamma_{e,k-1,\sigma^{\text{fdp}}} - 1 \\ \frac{1}{\Delta k}\rho^{\text{w}} A^{\text{cross}}_e L_e & \\ \vdots & \vdots \\ \frac{1}{\Delta k}\rho^{\text{w}} A^{\text{cross}}_e L_e & \text{for } \zeta = |\mathbb{S}_\zeta| \end{cases}$$

$$\forall\, e \in \mathbb{S}^{\text{pipe}}_e,\ \ k > 1 \in \mathbb{S}_k,\ \ \zeta \in \mathbb{S}_\zeta,\ \ \sigma^{\text{fdp}} \in \mathbb{S}_\sigma \tag{2.83}$$

The values of the predicted water masses $\widehat{\dot{m}}$ for the first position ζ_1 are updated with measurements or latest predictions resulting from the last optimization $\dot{m}$. Then all parameters which can be copied from the previous time step are utilized. In this context, Definition 2.2 has to be taken into account, as it concretizes the notation. For the second position $\zeta = 2$, the water mass from the previous time step $\widehat{\dot{m}}_{e,k-1,\zeta_\gamma,\sigma^{\text{fdp}}}$, is reduced by the amount of water, which has already left the pipeline. For this assignment the values of two further parameters γ and R^{wm} are necessary, see Figure 2.12. The exact form of calculation of these parameters is given in equation (2.98a) and (2.99a). The first describes the position of the last water mass which is leaving the pipeline, in the respective time step and flow direction. The second defines the aggregated water mass of all water elements on the positions from the beginning of the pipeline up to the one determined by γ. Note, that these parameters γ and R^{wm} are always updated based on the current values of the predicted mass flows $\widehat{\dot{m}}$. The last value, which can be copied is taken from the first position of the previous time step ζ_1. All further predicted water masses for the positions $\zeta > 2 + \gamma_{e,k-1,\sigma^{\text{fdp}}} - 1$ are then set to the value which fills the pipeline in a single time step $\frac{1}{\Delta k}\rho^{\text{w}} A^{\text{cross}}_e L_e$. Again, this prevents the unexpected case, that the pipeline could be “not completely filled” with water masses, within the model in case of a small number of positions $|\mathbb{S}_\zeta|$ and low mass flows in the current and past time steps[60]. Also take into account thc following aspects:

- In the special case, that the water mass R^{wm} is equal to the water mass of the filled pipeline, thus $R^{\mathrm{wm}} = \rho^{\mathrm{w}} A_e^{\mathrm{cross}} L_e$, the entire water mass $\widehat{m}_{e,k-1,\zeta_\gamma,\sigma^{\mathrm{fdp}}}$ is utilized for the new value of ζ_2, as no water of $\widehat{m}_{e,k-1,\zeta_\gamma,\sigma^{\mathrm{fdp}}}$ has left the pipeline in the previous time step.
- If the current (inserted) mass flow is large enough in relation to the pipeline length L, cross section A^{cross} and time step length Δk, the pipe is filled by the mass flow of a single time step. In this case the predicted mass flows for the positions $\zeta > 2$ still need to be copied, as these values affect the result of the outlet temperature of the pipeline T^{out2}.

Further, the rolling horizon time step shift for Case A.2 of the entering time step matrix $\boldsymbol{Z}$ is given by:

$$Z_{e,k,\zeta,\sigma^{\mathrm{fdp}},\tilde{k},\tilde{\sigma}} = \begin{cases} 1, \text{ if } \zeta = 1, \ \tilde{k} = k, \text{ and } \sigma^{\mathrm{fdp}} = \tilde{\sigma}, \\ \quad \text{or if } \zeta > 1, \text{ and } k > 1, \text{ and } Z_{e,k-1,\mathring{\zeta},\sigma^{\mathrm{fdp}},\tilde{k},\tilde{\sigma}} = 1 \\ \quad\quad \text{with } \mathring{\zeta} = 2 + \gamma_{e,k-1,\sigma^{\mathrm{fdp}}} - \zeta, \\ 0, \text{ else}, \end{cases}$$
$$\forall\, e \in \mathbb{S}_e^{\mathrm{pipe}}, \quad k, \tilde{k} \in \mathbb{S}_k, \quad \zeta, \mathring{\zeta} \in \mathbb{S}_\zeta, \quad \sigma^{\mathrm{fdp}}, \tilde{\sigma} \in \mathbb{S}_\sigma \tag{2.84}$$

Elements of the sparse matrix are nonzero if either the element represents a water mass which is just about to enter the pipeline or if the element represents a water mass which was in the pipeline before. In this case the conditions (considering the entering time step $\tilde{k}$ and entering flow direction $\tilde{\sigma}$) can be copied from the opposite flow direction which was present in the previous time step (on the FDP) in the regarded Case A.2[61,62].

Remark 2.3:

The index $\mathring{\zeta}$ is used here, to determine the positions of the last time step, which are now “copied” to the current time step. Thereby, the $\mathring{\square}$ operator is chosen, instead of the previously used $\tilde{\square}$ operator within the indices of e.g. $\boldsymbol{Z}$. This difference is chosen as the $\tilde{\square}$ operator refers to other time steps and flow directions, where the water masses entered the pipeline, while the $\mathring{\square}$ operator indicates a certain position which is determined by the copying procedure.

---

60 Theoretically, all predicted water masses $\widehat{m}$ for the positions $\zeta > 2 + \gamma_{e,k-1,\sigma^{\mathrm{fdp}}} - 1 + 1$ could also be set to 0 as they are not further needed for the approach in theory. However, defining them as in equation (2.83), also leads to feasible solutions of (2.98a) in the very unlikely case of previous and current mass flow values of $\widehat{m} \approx 0$ and numeric inaccuracies, e.g. if stop criteria of solvers would be loosened.

61 Note that σ^{fdp} is dependent on the time step here, as given in Definition 2.1 above.

62 The used values of γ are always updated, based on the new determined values of $\widehat{m}$ in the assignment process.

The elements of the past time step entering time matrix $\boldsymbol{Z}^{\mathrm{p}}$ is defined by:

$$Z^{\mathrm{p}}_{e,k,\zeta,\sigma^{\mathrm{fdp}},k^{\mathrm{p}},\tilde{\sigma}} = \begin{cases} 1, \text{ if } k = \tilde{k} = k^{\mathrm{p}} = 1, \text{ and } \mathring{\zeta}_1 = 2 + \gamma^{\mathrm{brh}}_{e,k_1,\sigma^{\mathrm{fdp}}} - \zeta, \\ \quad \text{and } Z^{\mathrm{brh}}_{e,k_1,\mathring{\zeta}_1,\sigma^{\mathrm{fdp}},\tilde{k}_1,\tilde{\sigma}} = 1, \text{and } \sigma^{\mathrm{fdp}}_{e,k} = \tilde{\sigma}_{e,k}, \\ \text{or if } k > 1, \text{ and } Z^{\mathrm{p}}_{e,k-1,\mathring{\zeta},\sigma^{\mathrm{fdp}},k^{\mathrm{p}},\tilde{\sigma}} = 1 \\ \quad \text{with } \mathring{\zeta} = 2 + \gamma_{e,k-1,\sigma^{\mathrm{fdp}}} - \zeta, \\ \text{or if } \zeta > 1, \; k = 1, \; k^{\mathrm{p}} > 1, \text{ and } Z^{\mathrm{p,brh}}_{e,k_1,\mathring{\zeta},\sigma^{\mathrm{fdp}},k^{\mathrm{p}}-1,\tilde{\sigma}} = 1, \\ \quad \text{with } \mathring{\zeta} = 2 + \gamma^{\mathrm{brh}}_{e,k_1,\sigma^{\mathrm{fdp}}} - \zeta, \\ 0, \text{ else}, \end{cases}$$

$$\forall\, e \in \mathbb{S}^{\mathrm{pipe}}_{e}, \;\; k,\tilde{k} \in \mathbb{S}_k, \;\; k^{\mathrm{p}} \in \mathbb{S}_{k^{\mathrm{p}}}, \;\; \zeta,\mathring{\zeta} \in \mathbb{S}_{\zeta}, \;\; \sigma^{\mathrm{fdp}},\tilde{\sigma} \in \mathbb{S}_{\sigma} \qquad (2.85)$$

For the following three cases the elements are nonzero:

- First, visually speaking, a nonzero element can be transferred from $\boldsymbol{Z}$ to $\boldsymbol{Z}^{\mathrm{p}}$ in the first time step k_1. This is the element which was in the first position ζ_1 at the first time step k_1 before the rolling horizon time step shift.
- The second condition for nonzero elements represents the *copy from previous time step* part for matrix elements of $\boldsymbol{Z}^{\mathrm{p}}$. In this matter, the mapping of the respective positions is achieved by the condition $\mathring{\zeta} = 2 + \gamma_{e,k-1,\sigma^{\mathrm{fdp}}} - \zeta$, which is represented by the arrows in Figure 2.15.
- The third and last case resulting in nonzero elements of matrix $\boldsymbol{Z}^{\mathrm{p}}$ enables to maintain the information of the earlier past time steps $k^{\mathrm{p}} > 1$ throughout the rolling horizon time step shift.

The following Example 2.1 illustrates the provided equations (2.84) and (2.85).

Example 2.1:

The regarded example is shown in Figure 2.17. As this is a visualization of ***Case A.2****, the flow directions looked at within this example represent the flow directions on the flow direction path* $\sigma^{\mathrm{fdp}}_{e_1,k_3} = \sigma^+$ *and* $\sigma^{\mathrm{fdp}}_{e_1,k_4} = \sigma^-$*. The flow directions not on the flow direction path are shown transparent in Figure 2.17.*

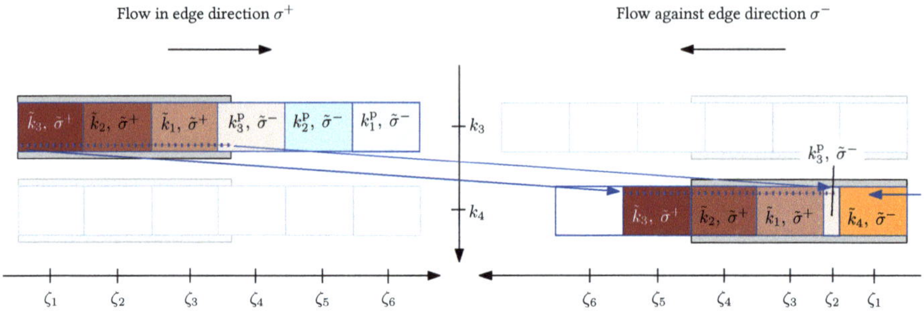

Figure 2.17: Example for Case A.2: Basic idea of approach if the flow direction on the flow direction path σ^{fdp} within a pipeline e changes between two consecutive time steps k and $k+1$.

For this example it is assumed, that the $\boldsymbol{Z}$ *and* $\boldsymbol{Z}^{\mathrm{p}}$ *are already updated for pipeline/edge* e_1*, time step* k_3 *with positive flow direction* σ^+*. These values are shown in Figure 2.17 and below:*

$$\boldsymbol{Z}_{e_1,k_3,\zeta,\sigma^+,\tilde{k},\tilde{\sigma}^+} = \begin{array}{c} \\ \zeta_1 \\ \zeta_2 \\ \zeta_3 \\ \zeta_4 \\ \zeta_5 \\ \zeta_6 \end{array} \begin{array}{c} \begin{array}{cccc} \tilde{k}_1 & \tilde{k}_2 & \tilde{k}_3 & \tilde{k}_4 \end{array} \\ \begin{bmatrix} 0 & 0 & 1 & 0 \\ 0 & 1 & 0 & 0 \\ 1 & 0 & 0 & 0 \\ 0 & 0 & 0 & 0 \\ 0 & 0 & 0 & 0 \\ 0 & 0 & 0 & 0 \end{bmatrix} \end{array}, \quad \boldsymbol{Z}_{e_1,k_3,\zeta,\sigma^+,\tilde{k},\tilde{\sigma}^-} = \begin{array}{c} \\ \zeta_1 \\ \zeta_2 \\ \zeta_3 \\ \zeta_4 \\ \zeta_5 \\ \zeta_6 \end{array} \begin{array}{c} \begin{array}{cccc} \tilde{k}_1 & \tilde{k}_2 & \tilde{k}_3 & \tilde{k}_4 \end{array} \\ \begin{bmatrix} 0 & 0 & 0 & 0 \\ 0 & 0 & 0 & 0 \\ 0 & 0 & 0 & 0 \\ 0 & 0 & 0 & 0 \\ 0 & 0 & 0 & 0 \\ 0 & 0 & 0 & 0 \end{bmatrix} \end{array} \tag{2.86}$$

$$\boldsymbol{Z}^{\mathrm{p}}_{e_1,k_3,\zeta,\sigma^+,k^{\mathrm{p}},\tilde{\sigma}^+} = \begin{array}{c} \\ \zeta_1 \\ \zeta_2 \\ \zeta_3 \\ \zeta_4 \\ \zeta_5 \\ \zeta_6 \end{array} \begin{array}{c} \begin{array}{cccc} k^{\mathrm{p}}_1 & k^{\mathrm{p}}_2 & k^{\mathrm{p}}_3 & k^{\mathrm{p}}_4 \end{array} \\ \begin{bmatrix} 0 & 0 & 0 & 0 \\ 0 & 0 & 0 & 0 \\ 0 & 0 & 0 & 0 \\ 0 & 0 & 0 & 0 \\ 0 & 0 & 0 & 0 \\ 0 & 0 & 0 & 0 \end{bmatrix} \end{array}, \quad \boldsymbol{Z}^{\mathrm{p}}_{e_1,k_3,\zeta,\sigma^+,k^{\mathrm{p}},\tilde{\sigma}^-} = \begin{array}{c} \\ \zeta_1 \\ \zeta_2 \\ \zeta_3 \\ \zeta_4 \\ \zeta_5 \\ \zeta_6 \end{array} \begin{array}{c} \begin{array}{cccc} k^{\mathrm{p}}_1 & k^{\mathrm{p}}_2 & k^{\mathrm{p}}_3 & k^{\mathrm{p}}_4 \end{array} \\ \begin{bmatrix} 0 & 0 & 0 & 0 \\ 0 & 0 & 0 & 0 \\ 0 & 0 & 0 & 0 \\ 0 & 0 & 1 & 0 \\ 0 & 1 & 0 & 0 \\ 1 & 0 & 0 & 0 \end{bmatrix} \end{array} \tag{2.87}$$

Based on the above, the novel matrix $\boldsymbol{Z}$ and $\boldsymbol{Z}^{\mathrm{p}}$ entries for the next time step k_4 with negative flow direction σ^- are calculated with (2.84) *as well as* (2.85) *and given by:*

$$\boldsymbol{Z}_{e_1,k_4,\zeta,\sigma^-,\tilde{k},\tilde{\sigma}^+} = \begin{array}{c} \\ \zeta_1 \\ \zeta_2 \\ \zeta_3 \\ \zeta_4 \\ \zeta_5 \\ \zeta_6 \end{array} \begin{array}{c} \begin{array}{cccc} \tilde{k}_1 & \tilde{k}_2 & \tilde{k}_3 & \tilde{k}_4 \end{array} \\ \begin{bmatrix} 0 & 0 & 0 & 0 \\ 0 & 0 & 0 & 0 \\ 1 & 0 & 0 & 0 \\ 0 & 1 & 0 & 0 \\ 0 & 0 & 1 & 0 \\ 0 & 0 & 1 & 0 \end{bmatrix} \end{array}, \quad \boldsymbol{Z}_{e_1,k_4,\zeta,\sigma^-,\tilde{k},\tilde{\sigma}^-} = \begin{array}{c} \\ \zeta_1 \\ \zeta_2 \\ \zeta_3 \\ \zeta_4 \\ \zeta_5 \\ \zeta_6 \end{array} \begin{array}{c} \begin{array}{cccc} \tilde{k}_1 & \tilde{k}_2 & \tilde{k}_3 & \tilde{k}_4 \end{array} \\ \begin{bmatrix} 0 & 0 & 0 & 1 \\ 0 & 0 & 0 & 0 \\ 0 & 0 & 0 & 0 \\ 0 & 0 & 0 & 0 \\ 0 & 0 & 0 & 0 \\ 0 & 0 & 0 & 0 \end{bmatrix} \end{array} \tag{2.88}$$

$$\boldsymbol{Z}^{\mathrm{p}}_{e_1,k_4,\zeta,\sigma^-,k^{\mathrm{p}},\tilde{\sigma}^+} = \begin{array}{c} \\ \zeta_1 \\ \zeta_2 \\ \zeta_3 \\ \zeta_4 \\ \zeta_5 \\ \zeta_6 \end{array} \begin{array}{c} \begin{array}{cccc} k^{\mathrm{p}}_1 & k^{\mathrm{p}}_2 & k^{\mathrm{p}}_3 & k^{\mathrm{p}}_4 \end{array} \\ \begin{bmatrix} 0 & 0 & 0 & 0 \\ 0 & 0 & 0 & 0 \\ 0 & 0 & 0 & 0 \\ 0 & 0 & 0 & 0 \\ 0 & 0 & 0 & 0 \\ 0 & 0 & 0 & 0 \end{bmatrix} \end{array}, \quad \boldsymbol{Z}^{\mathrm{p}}_{e_1,k_4,\zeta,\sigma^-,k^{\mathrm{p}},\tilde{\sigma}^-} = \begin{array}{c} \\ \zeta_1 \\ \zeta_2 \\ \zeta_3 \\ \zeta_4 \\ \zeta_5 \\ \zeta_6 \end{array} \begin{array}{c} \begin{array}{cccc} k^{\mathrm{p}}_1 & k^{\mathrm{p}}_2 & k^{\mathrm{p}}_3 & k^{\mathrm{p}}_4 \end{array} \\ \begin{bmatrix} 0 & 0 & 0 & 0 \\ 0 & 0 & 1 & 0 \\ 0 & 0 & 0 & 0 \\ 0 & 0 & 0 & 0 \\ 0 & 0 & 0 & 0 \\ 0 & 0 & 0 & 0 \end{bmatrix} \end{array} \tag{2.89}$$

Note, that the value $\gamma_{e,k-1,\sigma^{\mathrm{fdp}}}$ used in (2.84) *to determine the new values of $\boldsymbol{Z}_{e_1,k_4,\zeta,\sigma^-,\tilde{k},\tilde{\sigma}}$ is given by $\gamma_{e_1,k_3,\sigma^+} = 4$. Further, the same value is used in* (2.85) *to determine $\boldsymbol{Z}_{e_1,k_4,\zeta,\sigma^-,k^{\mathrm{p}},\tilde{\sigma}}$. The blue marked entry is optional and represents the additional accuracy measure described in Remark 2.4.*

The number of time steps a water mass was in the pipeline κ is updated for the first time step k_1 by:

$$\kappa_{e,k_1,\zeta,\sigma^{\mathrm{fdp}}} = \begin{cases} 0 & \text{for } \zeta = 1 \\ \kappa^{\mathrm{brh}}_{e,k_1,\mathring{\zeta}_\gamma,\sigma^{\mathrm{fdp}}} + 1 & \text{for } \zeta = 2 \\ \kappa^{\mathrm{brh}}_{e,k_1,\mathring{\zeta}_{\gamma-1},\sigma^{\mathrm{fdp}}} + 1 & \text{for } \zeta = 3 \\ \vdots & \vdots \\ \kappa^{\mathrm{brh}}_{e,k_1,\mathring{\zeta}_1,\sigma^{\mathrm{fdp}}} + 1 & \text{for } \zeta = 2 + \gamma^{\mathrm{brh}}_{e,k_1,\sigma^{\mathrm{fdp}}} - 1 \\ 0 & \\ \vdots & \vdots \\ 0 & \text{for } \zeta = |\mathbb{S}_\zeta| \end{cases}$$

$$\forall\, e \in \mathbb{S}^{\mathrm{pipe}}_e,\ \ \zeta, \mathring{\zeta} \in \mathbb{S}_\zeta,\ \ \tilde{\sigma} \in \mathbb{S}_\sigma \tag{2.90}$$

and for all following time steps by:

$$\kappa_{e,k,\zeta,\sigma^{\mathrm{fdp}}} = \begin{cases} 0 & \text{for } \zeta = 1 \\ \kappa_{e,k-1,\mathring{\zeta}_\gamma,\sigma^{\mathrm{fdp}}} + 1 & \text{for } \zeta = 2 \\ \kappa_{e,k-1,\mathring{\zeta}_{\gamma-1},\sigma^{\mathrm{fdp}}} + 1 & \text{for } \zeta = 3 \\ \vdots & \vdots \\ \kappa_{e,k-1,\mathring{\zeta}_1,\sigma^{\mathrm{fdp}}} + 1 & \text{for } \zeta = 2 + \gamma_{e,k-1,\sigma^{\mathrm{fdp}}} - 1 \\ 0 & \\ \vdots & \vdots \\ 0 & \text{for } \zeta = |\mathbb{S}_\zeta| \end{cases}$$

$$\forall\, e \in \mathbb{S}^{\mathrm{pipe}}_e,\ \ k > 1 \in \mathbb{S}_k,\ \ \zeta, \mathring{\zeta} \in \mathbb{S}_\zeta,\ \ \tilde{\sigma} \in \mathbb{S}_\sigma \tag{2.91}$$

As in equation (2.81), the elements in the first position $\zeta = 1$ of the pipeline are set to 0, while all further values are *copied from the previous time step* and increased by one. The values of all elements, which are not known as they can not be copied from the previous time step as $\zeta > 2 + \gamma_{e,k-1,\sigma^{\mathrm{fdp}}} - 1$, are set to $\kappa = 0$, as they also do not impact the calculation of the outlet temperature[63].

After all parameters on the rolling horizon have been newly assigned as given above, the parameters for all cases not on the flow direction path are updated based on the new values on the flow direction path as given below.

[63] In contrast to the assignment of the predicted mass flows $\widehat{m}$ in equation (2.83), no safety definition is performed here, as the value of κ has much less impact on T^{out2} as the values of $\widehat{m}$.

Case B: Rolling horizon time step shift of the pipeline parameters not on the flow direction path

In analogy to Case A.2 stated above, the central idea of Case B is to *insert and copy from the current time step*, as shown in Figure 2.18. For the first position ζ_1 measurement data or results of the last optimization are utilized for the parameter definition. Then, values of the parameters for the positions $\zeta > 1$ OFDP are copied from the parameters which have already been updated on the FDP. The main difference between Case B and Case A.2 is that parameters describing water masses of different parts of the "source pipeline" are copied in both cases. In Case A.2 the water masses copied from the previous time step are found in the positions $\zeta \leq \zeta_\gamma$, describing all water masses in the pipeline up to the last water mass which is leaving the pipeline in the previous time step. While in Case B all water masses starting from the second position ζ_2 up to the first water mass which is leaving the pipeline ζ_ε in the current time step are copied. Following this principle the predicted water masses $\widehat{m}$ are given as stated:

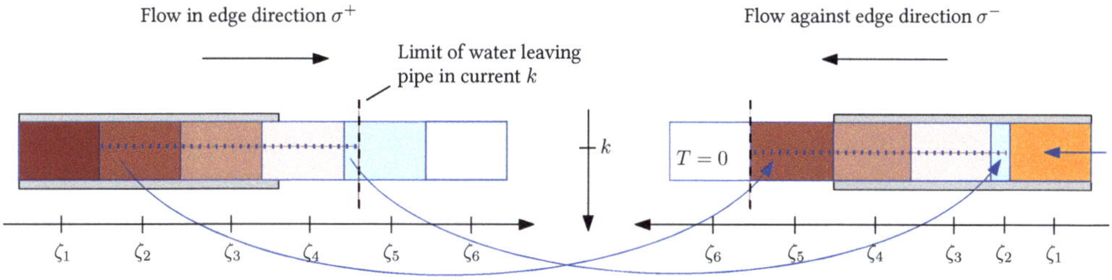

Figure 2.18: Case B: Basic idea of new assignment of the pipeline parameters opposed to the flow direction path σ^{ofdp} for a time step k by inserting and copying from the current time step. Note that in this example $\sigma^{\text{fdp}} = \sigma^+$ and thus $\sigma^{\text{ofdp}} = \sigma^-$. Further it should be kept in mind, that the amount of water flowing into the pipeline is equal to the amount of water leaving the pipeline at a certain time step, see Figure 2.12. An explanation of the water mass marked with $T = 0$ is provided later in Remark 2.4.

$$\widehat{\dot{m}}_{e,k,\zeta,\sigma^{\text{ofdp}}} = \begin{cases} |\dot{m}_{e,k}| & \text{for } \zeta = 1 \\ \widehat{m}_{e,k,\mathring{\zeta}_\varepsilon,\sigma^{\text{fdp}}} - \frac{1}{\Delta k}\left(S^{\text{wm}}_{e,k,\sigma^{\text{fdp}}} - \rho^{\text{w}} A^{\text{cross}}_e L_e\right) & \text{for } \zeta = 2 \\ \widehat{m}_{e,k,\mathring{\zeta}_{\varepsilon-1},\sigma^{\text{fdp}}} & \text{for } \zeta = 3 \\ \vdots & \vdots \\ \widehat{m}_{e,k,\mathring{\zeta}_2,\sigma^{\text{fdp}}} & \text{for } \zeta = 2 + \varepsilon_{e,k,\sigma^{\text{fdp}}} - 1 \\ \frac{1}{\Delta k}\rho^{\text{w}} A^{\text{cross}}_e L_e & \\ \vdots & \vdots \\ \frac{1}{\Delta k}\rho^{\text{w}} A^{\text{cross}}_e L_e & \text{for } \zeta = |\mathbb{S}_\zeta| \end{cases}$$

$$\forall\, e \in \mathbb{S}^{\text{pipe}}_e,\quad k \in \mathbb{S}_k,\quad \zeta, \tilde{\zeta} \in \mathbb{S}_\zeta,\quad \sigma^{\text{fdp}}, \sigma^{\text{ofdp}} \in \mathbb{S}_\sigma \tag{2.92}$$

Here, and in the following equations it is important to note that the flow directions are distinguished as on the FDP σ^{fdp} and OFDP σ^{ofdp}. The predicted mass flow of the first position ζ_1 is updated with the last values of the optimization or if available the respective sensors. Then, for the second position ζ_2, the parameter is updated by subtracting the water mass S^{wm}, see Figure 2.12, from the sum of the entire water mass of the pipeline $\rho^{\text{w}} A^{\text{cross}}_e L_e$ and the first water mass which is currently leaving the pipeline in the FDP case. This first water mass has the position ε. The parameters ε and S^{wm} are calculated with equation (2.98b) and (2.99b). Note, that ε and S^{wm} are always updated with the latest values of $\widehat{m}$. All following positions with $2 < \zeta \leq 2 + \varepsilon_{e,k,\sigma^{\mathrm{fdp}}} - 1$ are simply copied. In analogy to equation (2.83) the predicted water masses $\widehat{m}$ are set to the mass flow, which is necessary to fill the entire pipeline within one time step $\frac{1}{\Delta k}\rho^{\mathrm{w}} A_e^{\mathrm{cross}} L_e$ in order to prevent model inaccuracies in case of numeric issues[60].

The elements of the entering time step and flow direction matrix $\boldsymbol{Z}$ are given as follows for Case B:

$$Z_{e,k,\zeta,\sigma^{\mathrm{ofdp}},\tilde{k},\tilde{\sigma}} = \begin{cases} 1, \text{ if } \zeta = 1 \text{ and } \tilde{k} = k \text{ and } \sigma^{\mathrm{ofdp}} = \tilde{\sigma}, \\ \quad \text{or if } Z_{e,k,\mathring{\zeta},\sigma^{\mathrm{fdp}},\tilde{k},\tilde{\sigma}} = 1 \\ \quad\quad \text{with } \mathring{\zeta} = 2 + \varepsilon_{e,k,\sigma^{\mathrm{fdp}}} - \zeta, \\ 0, \text{ else}, \end{cases}$$

$$\forall\, e \in \mathbb{S}_e^{\mathrm{pipe}},\;\; k, \tilde{k} \in \mathbb{S}_k,\;\; \zeta, \mathring{\zeta} \in \mathbb{S}_\zeta,\;\; \sigma^{\mathrm{fdp}}, \sigma^{\mathrm{ofdp}}, \tilde{\sigma} \in \mathbb{S}_\sigma \qquad (2.93)$$

Therein, the elements are nonzero if either a new element is *inserted* or if the elements are *copied* from the FDP of *the current time step.* In this context, the equation $\mathring{\zeta} = 2 + \varepsilon_{e,k,\sigma^{\mathrm{fdp}}} - \zeta$ maps the respective positions of the water mass elements in the pipeline model on the FDP and OFDP. An example for the determination of $\boldsymbol{Z}$ in Case B is provided in Example 2.2. The past entering time step and flow direction matrix $\boldsymbol{Z}^{\mathrm{p}}$ is defined similarly as:

$$Z^{\mathrm{p}}_{e,k,\zeta,\sigma^{\mathrm{ofdp}},k^{\mathrm{p}},\tilde{\sigma}} = \begin{cases} 1, \text{ if } Z^{\mathrm{p}}_{e,k,\tilde{\zeta},\sigma^{\mathrm{fdp}},k^{\mathrm{p}},\tilde{\sigma}} = 1 \\ \quad\quad \text{with } \tilde{\zeta} = 2 + \varepsilon_{e,k,\sigma^{\mathrm{fdp}}} - \zeta, \\ 0, \text{ else}, \end{cases}$$

$$\forall\, e \in \mathbb{S}_e^{\mathrm{pipe}},\;\; k, \tilde{k} \in \mathbb{S}_k,\;\; k^{\mathrm{p}} \in \mathbb{S}_{k^{\mathrm{p}}},\;\; \zeta, \tilde{\zeta} \in \mathbb{S}_\zeta,\;\; \sigma^{\mathrm{fdp}}, \sigma^{\mathrm{ofdp}}, \tilde{\sigma} \in \mathbb{S}_\sigma \qquad (2.94)$$

Note, that this updating process is simple here as only some information is *copied* from the FDP of *the current time step* therefore. Exemplary values are also given in the following Example 2.2.

Example 2.2:

*An example for **Case B** is depicted in Figure 2.19. Note, that the flow direction on the flow direction path is* $\sigma^{\mathrm{fdp}}_{e_1,k_3} = \sigma^{+}$.

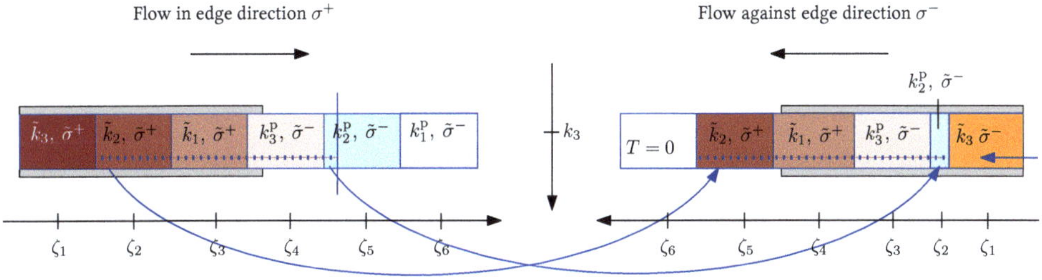

Figure 2.19: Example for Case B: Basic idea of new assignment of the pipeline parameters opposed to the flow direction path σ^{ofdp} for time step k_3 by inserting and copying from the current time step. Note that in this example $\sigma^{\mathrm{fdp}} = \sigma^{+}$ and thus $\sigma^{\mathrm{ofdp}} = \sigma^{-}$. An explanation of the water mass marked with $T = 0$ is provided in Remark 2.4.

For this example it is assumed, that the $\boldsymbol{Z}$ *and* $\boldsymbol{Z}^{\mathrm{p}}$ *are already updated for pipeline/edge* e_1, *time step* k_3 *with positive flow direction* σ^{+}. *These values are shown in Figure 2.19 and given in* (2.86) *and* (2.87). *Based on these, the novel matrix entries of* $\boldsymbol{Z}$ *and* $\boldsymbol{Z}^{\mathrm{p}}$ *for the opposed negative flow direction* σ^{-} *are calculated with* (2.93) *as well as* (2.94) *and given by:*

$$\boldsymbol{Z}_{e_1,k_3,\zeta,\sigma^-,\tilde{k},\tilde{\sigma}^+} = \begin{array}{c} \\ \zeta_1 \\ \zeta_2 \\ \zeta_3 \\ \zeta_4 \\ \zeta_5 \\ \zeta_6 \end{array} \begin{array}{c} \begin{array}{cccc} \tilde{k}_1 & \tilde{k}_2 & \tilde{k}_3 & \tilde{k}_4 \end{array} \\ \begin{bmatrix} 0 & 0 & 0 & 0 \\ 0 & 0 & 0 & 0 \\ 0 & 0 & 0 & 0 \\ 1 & 0 & 0 & 0 \\ 0 & 1 & 0 & 0 \\ 0 & {\color{blue}1} & 0 & 0 \end{bmatrix} \end{array}, \quad \boldsymbol{Z}_{e_1,k_3,\zeta,\sigma^-,\tilde{k},\tilde{\sigma}^-} = \begin{array}{c} \\ \zeta_1 \\ \zeta_2 \\ \zeta_3 \\ \zeta_4 \\ \zeta_5 \\ \zeta_6 \end{array} \begin{array}{c} \begin{array}{cccc} \tilde{k}_1 & \tilde{k}_2 & \tilde{k}_3 & \tilde{k}_4 \end{array} \\ \begin{bmatrix} 0 & 0 & 1 & 0 \\ 0 & 0 & 0 & 0 \\ 0 & 0 & 0 & 0 \\ 0 & 0 & 0 & 0 \\ 0 & 0 & 0 & 0 \\ 0 & 0 & 0 & 0 \end{bmatrix} \end{array} \tag{2.95}$$

$$\boldsymbol{Z}^{\mathrm{p}}_{e_1,k_3,\zeta,\sigma^-,k^{\mathrm{p}},\tilde{\sigma}^+} = \begin{array}{c} \\ \zeta_1 \\ \zeta_2 \\ \zeta_3 \\ \zeta_4 \\ \zeta_5 \\ \zeta_6 \end{array} \begin{array}{c} \begin{array}{cccc} k^{\mathrm{p}}_1 & k^{\mathrm{p}}_2 & k^{\mathrm{p}}_3 & k^{\mathrm{p}}_4 \end{array} \\ \begin{bmatrix} 0 & 0 & 0 & 0 \\ 0 & 0 & 0 & 0 \\ 0 & 0 & 0 & 0 \\ 0 & 0 & 0 & 0 \\ 0 & 0 & 0 & 0 \\ 0 & 0 & 0 & 0 \end{bmatrix} \end{array}, \quad \boldsymbol{Z}^{\mathrm{p}}_{e_1,k_3,\zeta,\sigma^-,k^{\mathrm{p}},\tilde{\sigma}^-} = \begin{array}{c} \\ \zeta_1 \\ \zeta_2 \\ \zeta_3 \\ \zeta_4 \\ \zeta_5 \\ \zeta_6 \end{array} \begin{array}{c} \begin{array}{cccc} k^{\mathrm{p}}_1 & k^{\mathrm{p}}_2 & k^{\mathrm{p}}_3 & k^{\mathrm{p}}_4 \end{array} \\ \begin{bmatrix} 0 & 0 & 0 & 0 \\ 0 & 1 & 0 & 0 \\ 0 & 0 & 1 & 0 \\ 0 & 0 & 0 & 0 \\ 0 & 0 & 0 & 0 \\ 0 & 0 & 0 & 0 \end{bmatrix} \end{array} \tag{2.96}$$

The value $\varepsilon_{e,k,\sigma^{\mathrm{fdp}}}$ *used in* (2.93) *to determine the new values of* $\boldsymbol{Z}_{e_1,k_3,\zeta,\sigma^-,\tilde{k},\tilde{\sigma}}$ *is given by* $\varepsilon_{e_1,k_3,\sigma^+} = 5$. *Further, the same value is used in* (2.94) *to determine* $\boldsymbol{Z}_{e_1,k_3,\zeta,\sigma^-,k^{\mathrm{p}},\tilde{\sigma}}$. *The blue marked entry is optional and represents the additional accuracy measure described in Remark 2.4.*

Also the values for the number of time steps a water mass was in the pipeline κ are updated accordingly to the previous assignments in Case B:

$$\kappa_{e,k,\zeta,\sigma^{\text{ofdp}}} = \begin{cases} 0 & \text{for } \zeta = 1 \\ \kappa_{e,k,\tilde{\zeta}_\varepsilon,\sigma^{\text{fdp}}} & \text{for } \zeta = 2 \\ \kappa_{e,k,\tilde{\zeta}_{\varepsilon-1},\sigma^{\text{fdp}}} & \text{for } \zeta = 3 \\ \vdots & \vdots \\ \kappa_{e,k,\tilde{\zeta}_2,\sigma^{\text{fdp}}} & \text{for } \zeta = 2 + \varepsilon_{e,k,\sigma^{\text{fdp}}} - 1 \\ 0 & \\ \vdots & \vdots \\ 0 & \text{for } \zeta = |\mathbb{S}_\zeta| \end{cases}$$

$$\forall\, e \in \mathbb{S}_e^{\text{pipe}},\ \ k \in \mathbb{S}_k,\ \ \zeta, \tilde{\zeta} \in \mathbb{S}_\zeta,\ \ \sigma^{\text{fdp}}, \sigma^{\text{ofdp}}, \tilde{\sigma} \in \mathbb{S}_\sigma \tag{2.97}$$

The values of κ for the new *inserted* water masses on position $\zeta = 1$ are set to $\kappa = 0$, while all further values are *copied* as in the earlier assignments of Case B above.

Remark 2.4:

When copying water masses from the current or previous time step in Case A.2 or Case B some water masses outside the newly assigned pipeline, will possess the temperature value $T = 0$. An example of this case is shown as the white water mass in Figure 2.18. In theory, the approach will always provide sufficiently filled pipes as well as out flowing water masses with accordingly assigned node temperatures to enable an accurate calculation of the outlet temperature of the pipelines T^{out2}. However, in case of numerical inaccuracy of the calculation of the outlet temperature of the pipeline, a small fraction of the last water mass with $T = 0$ could be used by the approach. The resulting error through this effect is neglectable. Still, for example if larger solver tolerances are to be used, this effect can be simply reduced. Therefore, visually speaking, the temperature is kept constant at the end of the pipeline, as soon as no temperature values are known any more. This is done by setting the last water masses of the pipeline model, which would usually not be set (thus $T = 0$), to the temperature of the last known water mass temperature. This would be the dark red water mass to the right of the white water mass in Figure 2.18. In this case all values of $\boldsymbol{Z}$ and $\boldsymbol{Z}^{\text{p}}$, which do not have a nonzero entry, over all dimensions for a certain position $\zeta \in \mathbb{S}_\zeta$, are set to the last set time step k or k^{p}. See Example 2.1 for an illustration.

In summary, the above explanations described how the pipeline parameters $Z_{e,k,\zeta,\sigma,\tilde{k},\tilde{\sigma}}$, $Z^{\text{p}}_{e,k,\zeta,\sigma,k^{\text{p}},\tilde{\sigma}}$, $\widehat{m}_{e,k,\zeta,\sigma}$, and $\kappa_{e,k,\zeta,\sigma}$ are updated the realize the rolling horizon time shift. The matrices $\boldsymbol{Z}$ and $\boldsymbol{Z}^{\text{p}}$ are used to assign the respective (past) node temperatures to the water masses in the pipelines. As explained in the upcoming paragraph the mass flow values $\widehat{m}$ are used to determine the weights w used to define which out flowing water mass should be taken into account by which percentage in the weighted sum used to calculate the lossless outlet temperature in (2.70). Further, the parameters κ, which describe the amounts of time steps a water mass was in the given pipeline, are used to calculate the length of stay of the leaving water mass.

Calculation of Pipeline Parameters Dependent on Updated Values of $\dot{m}$ and κ of the Water Masses Based on the newly calculated stored mass flow values $\widehat{\dot{m}}_{e,k,\zeta,\sigma}$, the pipeline position parameters $\gamma_{e,k,\sigma}$, $\varepsilon_{e,k,\sigma}$ are updated as defined below:

$$\gamma_{e,k,\sigma} = \min_{x} \left\{ x : \quad \sum_{\zeta=1}^{x} \left(\widehat{\dot{m}}_{e,k,\zeta,\sigma} \Delta k \right) \geq \rho^{\mathrm{w}} A_e^{\mathrm{cross}} L_e, \quad x \geq 1,\ x \in \mathbb{Z} \right\}, \quad \forall\, e \in \mathbb{S}_e^{\mathrm{pipe}}, k \in \mathbb{S}_k, \sigma \in \mathbb{S}_\sigma \tag{2.98a}$$

$$\varepsilon_{e,k,\sigma} = \min_{z} \left\{ z : \quad \sum_{\zeta=2}^{z} \left(\widehat{\dot{m}}_{e,k,\zeta,\sigma} \Delta k \right) \geq \rho^{\mathrm{w}} A_e^{\mathrm{cross}} L_e, \quad z \geq 2,\ z \in \mathbb{Z} \right\}, \quad \forall\, e \in \mathbb{S}_e^{\mathrm{pipe}}, k \in \mathbb{S}_k, \sigma \in \mathbb{S}_\sigma \tag{2.98b}$$

An intuitive understanding of the positions defined by these parameters can be obtained from Figure 2.12. The pipeline position parameter γ defines the temporally seen last water mass leaving the pipeline e in the current time step k for a given flow direction σ. Analogously the position ε defines the first water mass leaving the pipeline. Based on these values also the auxiliary water masses $R^{\mathrm{wm}}_{e,k,\sigma}$, and $S^{\mathrm{wm}}_{e,k,\sigma}$ can be calculated as stated:

$$R^{\mathrm{wm}}_{e,k,\sigma} = \sum_{\zeta=1}^{\gamma_{e,k,\sigma}} \left(\widehat{\dot{m}}_{e,k,\zeta,\sigma} \Delta k \right), \quad \forall\, e \in \mathbb{S}_e^{\mathrm{pipe}}, k \in \mathbb{S}_k, \sigma \in \mathbb{S}_\sigma \tag{2.99a}$$

$$S^{\mathrm{wm}}_{e,k,\sigma} = \begin{cases} \displaystyle\sum_{\zeta=1}^{\varepsilon_{e,k,\sigma}-1} \left(\widehat{\dot{m}}_{e,k,\zeta,\sigma} \Delta k \right), & \text{if } \varepsilon_{e,k,\sigma} \geq \gamma_{e,k,\sigma} + 1 \\ R^{\mathrm{wm}}_{e,k,\sigma}, & \text{else,} \end{cases} \quad \forall\, e \in \mathbb{S}_e^{\mathrm{pipe}}, k \in \mathbb{S}_k, \sigma \in \mathbb{S}_\sigma \tag{2.99b}$$

Regarding Figure 2.12 it is understood, that R^{wm} defines the aggregated water mass of all water elements on the positions from the beginning of the pipeline up to the one determined by γ. Similarly the auxiliary water mass S^{wm} is defined by the position $\varepsilon - 1$. Taking into account the parameters $\gamma_{e,k,\sigma}$, $\varepsilon_{e,k,\sigma}$, $R^{\text{wm}}_{e,k,\sigma}$, and $S^{\text{wm}}_{e,k,\sigma}$ also the weights can be defined $w_{e,k,\zeta,\sigma}$ as:

$$w_{e,k,\zeta,\sigma} = \frac{1}{|\dot{m}_{e,k}|\Delta k}(R^{\text{wm}}_{e,k,\sigma} - \rho^{\text{w}} A^{\text{cross}}_{e} L_e),$$
$$\text{if } \zeta = \gamma_{e,k,\sigma},\ \forall\, e \in \mathbb{S}^{\text{pipe}}_{e}, k \in \mathbb{S}_k, \sigma \in \mathbb{S}_\sigma, \tag{2.100}$$

$$w_{e,k,\zeta,\sigma} = \frac{1}{|\dot{m}_{e,k}|\Delta k}(|\widehat{\dot{m}}_{e,k,\zeta,\sigma}|\Delta k)$$
$$\text{if } \gamma_{e,k,\sigma} < \zeta < \varepsilon_{e,k,\sigma},\ \forall\, e \in \mathbb{S}^{\text{pipe}}_{e}, k \in \mathbb{S}_k, \sigma \in \mathbb{S}_\sigma \tag{2.101}$$

$$w_{e,k,\zeta,\sigma} = \begin{cases} \frac{1}{|\dot{m}_{e,k}|\Delta k}(|\dot{m}_{e,k}|\Delta k + \rho^{\text{w}} A^{\text{cross}}_{e} L_e - S^{\text{wm}}_{e,k,\sigma}), \\ \text{if } \zeta = \varepsilon_{e,k,\sigma} \text{ and } \gamma_{e,k,\sigma} < \varepsilon_{e,k,\sigma},\ \forall\, e \in \mathbb{S}^{\text{pipe}}_{e}, k \in \mathbb{S}_k, \sigma \in \mathbb{S}_\sigma, \\ \frac{1}{|\dot{m}_{e,k}|\Delta k}(|\dot{m}_{e,k}|\Delta k + \rho^{\text{w}} A^{\text{cross}}_{e} L_e - S^{\text{wm}}_{e,k,\sigma}) + w_{e,k,\zeta,\sigma}, \\ \text{if } \zeta = \varepsilon_{e,k,\sigma} \text{ and } \gamma_{e,k,\sigma} = \varepsilon_{e,k,\sigma},\ \forall\, e \in \mathbb{S}^{\text{pipe}}_{e}, k \in \mathbb{S}_k, \sigma \in \mathbb{S}_\sigma \end{cases} \tag{2.102}$$

Using these weights w, the length of stay within the pipeline $t_{e,k,\sigma}$ of the leaving water mass can be calculated by a weighted sum, as:

$$\begin{aligned} t_{e,k,\sigma} &= \big(w_{e,k,\gamma,\sigma}\kappa_{e,k,\gamma,\sigma} + \sum_{\zeta=\gamma+1}^{\varepsilon-1} w_{e,k,\zeta,\sigma}\kappa_{e,k,\zeta,\sigma} + w_{e,k,\varepsilon,\sigma}\kappa_{e,k,\varepsilon,\sigma}\big)\Delta k \\ &= \big(\sum_{\zeta=\gamma}^{\varepsilon} w_{e,k,\zeta,\sigma}\kappa_{e,k,\zeta,\sigma}\big)\Delta k, \qquad \forall\, e \in \mathbb{S}^{\text{pipe}}_{e}, k \in \mathbb{S}_k, \sigma \in \mathbb{S}_\sigma \end{aligned} \tag{2.103}$$

Therein the values of κ represent the amounts of time steps a certain water mass was in the given pipeline.

Based on the newly calculated weights w and the length of stay t, all parameters are given to calculate the lossless outlet temperature of a pipeline T^{out1} with (2.70) as well as the lossy outlet temperature T^{out2} in (2.71). Thereby, the entire thermal pipeline model is provided.

Upper Bound of Possible Modeling Errors This section determines an upper bound of possibly occurring errors from the above described thermal pipeline model. The accuracy of the presented model is dependent on the quality of the predicted mass flow estimates $\widehat{\dot{m}}$. These predicted values are then used to calculate the weights w, which then determine the length of stay of the water masses in the pipeline t and the outlet temperature T^{out2}. In case of perfect flow predictions, the presented model accuracy equals the results of the NM, showing very small RMSEs of 0.507 °C [MRMH21], as stated above. In case the predictions show deviations from the actual flow conditions within the DHN, an upper bound of the possible modeling error $\Delta T^{\text{max,dev}}_{e,k}$ for a given pipeline e and time step k can be derived as follows. Therefore, the maximal deviation of the actual temperature measured at the outlet of a pipeline T^{meas} and the results of the simulation based on the flow direction $T^{\text{out2}}_{e,k,\sigma+}$ or $T^{\text{out2}}_{e,k,\sigma-}$ is used. Next, T^{out2} is replaced by the possible minimal or maximal temperature values, which can occur at the outlet of the pipeline. These are the minimal or maximal values of all past or current node temperatures $T_{i,k^{\text{p}}}$ or $T_{i,k_{\text{al}}}$ of all nodes i bordering the respective edge e, thus fulfilling the condition $|A^{\mathrm{dhn}}_{i,e}| > 0$. Note, that the minimal outlet temperature values are calculated by taking into account the largest possible losses over the pipeline, thus using the longest possible length of stay of a water mass $t^{\max}_{e,k}$. The formal expression is:

$$
\begin{aligned}
\Delta T^{\max,\mathrm{dev}}_{e,k} &= \max \left(|T^{\mathrm{meas}}_{e,k} - T^{\mathrm{out2}}_{e,k,\sigma^+}|, |T^{\mathrm{meas}}_{e,k} - T^{\mathrm{out2}}_{e,k,\sigma^-}| \right) \\
&\leq \max_{\substack{\forall\, i \in \mathbb{S}^{\mathrm{dhn}}_i \\ \text{with } |A^{\mathrm{dhn}}_{i,e}|>0, \\ \tilde{k} \in \mathbb{S}_k, \\ k^{\mathrm{P}} \in \mathbb{S}_{k^{\mathrm{P}}}}} \Bigg(|T^{\mathrm{meas}}_{e,k} - T_{i,\tilde{k}}|, |T^{\mathrm{meas}}_{e,k} - T_{i,k^{\mathrm{P}}}|, \\
&\qquad |T^{\mathrm{meas}}_{e,k} - (T^{\mathrm{a}} + (T_{i,\tilde{k}} - T^{\mathrm{a}}) \exp(-\frac{t^{\max}_{e,k}}{\tau_e}))|, \\
&\qquad |T^{\mathrm{meas}}_{e,k} - (T^{\mathrm{a}} + (T_{i,k^{\mathrm{P}}} - T^{\mathrm{a}}) \exp(-\frac{t^{\max}_{e,k}}{\tau_e}))| \Bigg), \\
&\forall\, e \in \mathbb{S}^{\mathrm{pipe}}_e, k \in \mathbb{S}_k
\end{aligned}
\tag{2.104}
$$

The upper bound of possible modeling errors, can be approximated by the product of the time interval and the maximal gradient of temperature changes at nodes in DHN operation which are in the range of 3 °C/h [MMW+20] and 2 °C/min [Rat19, p.25-26]. In general DHN operators are interested in keeping these gradients small to prevent fast aging of the pipe insulation material [MMW+20]. When carrying out parameter studies to evaluate the model accuracy, the quality of the mass flow predictions is of high interest, as they directly impact the outlet temperature of the pipeline. These predicted mass flow values originate form the last optimization of the rolling horizon approach. Thus, in [Tsc19] a comparison of the lossy pipeline outlet temperatures, based on the predicted mass flow values, with the exact mass flow values for a small DHN with fixed mass flow directions was performed. The predicted mass flows were obtained from the rolling horizon optimization of the last time step, while the exact mass flow values resulted from the mass flows obtained from the current optimization. Based on this, it can be seen, that the amount of time steps included in the prediction horizon largely impact the accuracy of the predicted mass flow values and thus the outlet temperature. For a prediction horizon of 1 time step, see Figure 2.20, the deviations are below 4 K within the entire DHN except for a single time step, where the maximum deviation of 10.22 K is reached at Node 8. For a larger prediction horizon of 14 time steps, see Figure 2.20, the maximum error falls below 1.5 K and even 0.1 K after the stronger deviations created through the initialization of the rolling horizon approach. Further, in [SWW+19] it is reported, that mass flow predictions with maximum errors in the mean values of 1 % and 6 % in the standard deviation can be obtained. All in all, the given upper bound, the parameter study and the cited literature indicate that temperature deviations will most likely be in the lower single digit range. Further, the presented simulation results show the salient features of the combination of the presented thermal pipeline model with the rolling horizon approach, which will be presented in detail in Section 3.3. Through the rolling horizon approach, the quality of the predicted mass flow values will most likely increase toward the current operation time for a sufficient prediction horizon. Besides, within this model no additional numeric effects, as dispersive and dissipative errors, cause additional deviations as this is the case e.g. for finite difference DHN pipeline modeling approaches [MRMH21].

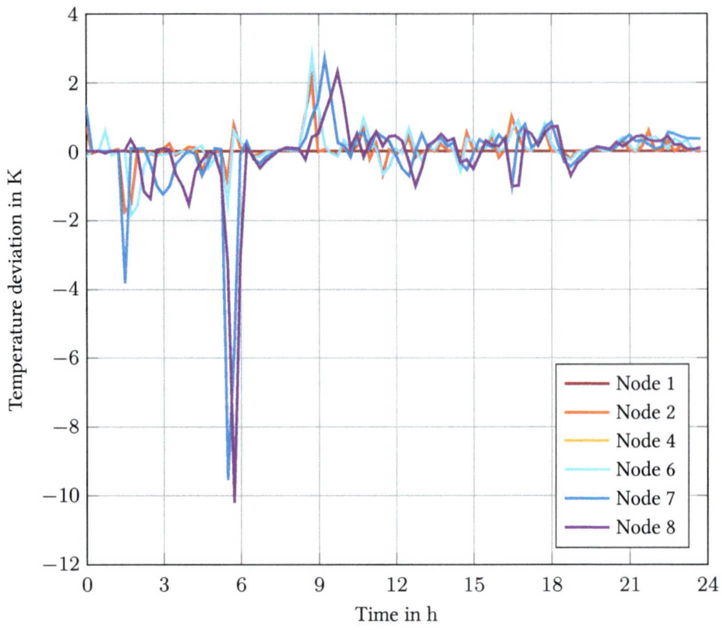

Figure 2.20: Deviation between the exact simulated temperature and the one calculated from the proposed pipeline model for 1 time step on the prediction horizon.

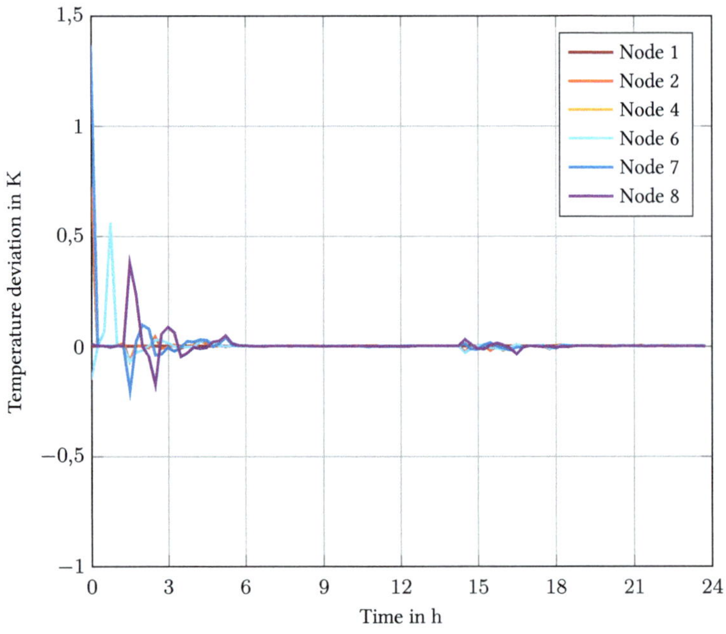

Figure 2.21: Deviation between the exact simulated temperature and the one calculated from the proposed pipeline model for 14 time steps on the prediction horizon.

2.4 Summary and Discussion

CEPDHNs are complex large systems that result in extensive network models. The model presented in this chapter can be represented as a nonlinear system of algebraic equations:

$$\boldsymbol{h}(\boldsymbol{x}) = \boldsymbol{0} \tag{2.105}$$

For that matter (2.105) is constructed by stacking the presented equations of all CEPDHN components for all time steps. Precisely, these equations are (2.11), (2.12), (2.15), (2.16), (2.17), (2.21), (2.23), (2.24), (2.26), (2.28), (2.29), (2.30), (2.31), (2.32), (2.33), (2.34), (2.35), (2.36), (2.39), (2.42), (2.43), (2.44), (2.45), (2.47), (2.49), (2.51), (2.52), (2.53), (2.54), (2.56), (2.57), (2.58), (2.59) , (2.60), (2.61), (2.62), (2.63), (2.64), (2.65), (2.66), (2.70), (2.71), (2.72), and (2.73).

Before this system (2.105) can be simulated/solved, based on an initial value $\boldsymbol{x}_0$, several input parameters need to be provided. These are the network parameters, the power injections or demand of all NPs, the set points of pumps and valves, and the parameters needed for the pipeline model. This shows, that in general the proposed model could also be used for simulating CEPDHNs. Still, this model was especially developed for application within operational optimization. This new CEPDHN model incorporates difficult aspects which are usually neglected within operation optimization of CEPDHNs, as: VFVT operation of DHNs including the effects of pipeline storage, VMFDs, control paths of pumps and differential pressure regulators, pipeline friction factor calculations based on close approximations of the Prandtl-Colebrook equation, and the full AC power flow.

As it can be seen from literature cited with the proper network element models, the presented high detail CEPDHN model is the result of a customized combination and extension of existing models. The most relevant works, with the largest impact on the provided network representation (2.105) are [Trö99, Opp15]. Note, that the model in [Opp15] was designed for simulation purposes. Thus, the model is not continuously differentiable as it relies on conditions, which need to be checked during the solution process. Further, this work does not incorporate EPNs and is based on the assumption of fixed flow directions. The model in [Trö99] neglects control paths and differential pressure regulators, and the EPN is modeled as a single bus. Also both approaches [Trö99, Opp15] do not provide TESS models, and use deviating thermal pipeline models. The contributed thermal pipeline model, first presented in this work, is built upon the node method which was first presented in [Ben91]. Combining this with ideas from [LWS+15] and [Trö99] as well as further improvements of the author, as stated in Section 2.3.3, led to new thermal pipeline model, which enables to take VMFDs into account. Further, as shown towards the end of Section 2.3.3 the thermal pipeline model profits from the operation principle of a rolling horizon approach, as the predicted mass flow values used to calculate the output temperatures tend to become more and more precise the closer the regarded time step is to the actual point in time. This is of importance as the propagation of the temperature fronts through the DHN pipelines are the only physical effect, that needs to be modeled through dynamic models, in the context of operational optimization [ZWW+21]. In total this CEPDHN model is the central element of the model based control approach of the TCS presented in the following chapter.

The comparatively high modeling detail of the presented CEPDHN model has two salient features. First, it enables a technically efficient form of CEPDHN operation. This is achieved by representing variable flows and variable temperature operating conditions, which are the core of the most efficient form of DHN operation [DLS+19, ZZZW18a]. Also the consideration of VMFDs, pipeline storage and the losses occurring in EPNs and DHNs account for technical efficiency, when considered within optimal operation approaches. Second, the aforementioned modeling detail has a direct effect on the possible solution set of market-based operation approaches, and thus on the maximum achievable welfare result. Since, for example preventing VMFDs can have severe consequences on the objective value/market outcome, see Section 1.2.5. In summary, the detailed CEPDHN model forms Contribution 2 of this work, bridging the element 1. a) of the research gap stated in Section 1.2.5.

3 Transactive Control System Design

This chapter outlines the specific methods used to design the TCS for CEPDHNs which is the central contribution of this work. Therefore the general system design is presented in Section 3.1. Therein, the different entities which are part of the presented approach are stated and their actions and information exchange are outlaid. This comprises the basic market and control mechanisms applied by design. A detailed description of possible and the selected market mechanisms is than given in Section 3.2. The subsequent Section 3.3 describes the control mechanisms, specially designed for the novel TCS. Parts of this chapter have been published in [MTK+22], [MGR+21], and [MIJSH22].

3.1 System Design

Within this section, the big picture and basic functionality of the new TCS system design, proposed in this work, is introduced. Note, that parts of the descriptions in this section have been previously published in [MTK+22]. For the optimal economic operation of CEPDHNs, FNPs should be operated in a way that maximizes social welfare [ZXL+17]. To this effect, an intraday TCS auction market[64], activated after settlement of a day-ahead energy market, is assumed[65] to coordinate the interests of all FNPs. The proposed intraday energy market is auction based and considered within the Independent System Operator Energy Management Systems (ISOEMSs), which are designed using a *distributed rolling horizon approach*, described in detail in Section 3.3. Therefore, the ISOEMSs take into account all necessary information as the nonlinear dynamic and stationary network models, operational constraints, predictions of power supply and demand of inflexible units, and bids of FNPs to determine the optimal control values/dispatch solutions that maximizes social welfare and ensures secure network operations[8] at the same time. Note, that the ISOEMSs therefore solve a nonlinear optimization problem in a parallel but coordinated manner in order to enable real-time implementation of this large scale dispatch problem. This indicates the necessity of a prior decomposition of the entire market clearing dispatch problem into multiple subproblems. As will be discussed in detail in Section 3.3.2, this decomposition is performed by decomposing the CEPDHN network into multiple operation zones[66]. Note, that the dispatch solution of the distributed ISOEMSs is the same as it would be calculated by a single ISOEMS for the entire CEPDHN. Thereby, the proposed market and control mechanisms, which are briefly presented in this section and discussed in detail in the following Sections 3.2 and 3.3, are integrated into a new TCS to

64 Precisely, this market represents the integration of an electric intraday market and a heat intraday market for every DHN, which are all cleared in a coordinated joint simultaneous procedure, as explained in detail in Section 3.2.

65 See Assumption 3.3 for more details.

66 The term *operation* zone is utilized here in order to enable to distinguish between operation and price zones in the following.

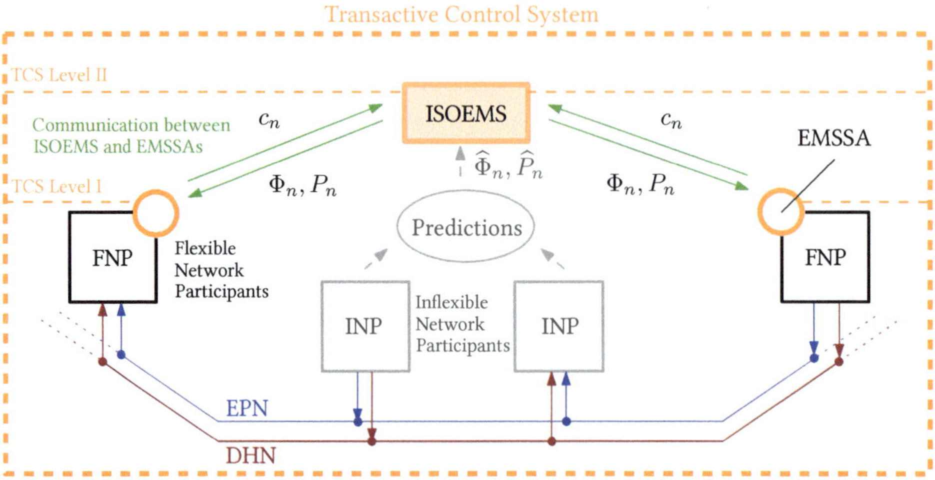

Figure 3.1: Proposed approach showing information flow between the two hierarchy levels of the TCS, i.e., the ISOEMS (TCS Level II) and the EMSSAs of the FNPs (TCS Level I), and the energy supply and demand of flexible and inflexible network participants connected to the CEPDHN.

optimally operate CEPDHNs. An overview of the proposed TCS is shown in Figure 3.1 for the case of a single ISOEMS, and in Figure 3.2 for multiple distributed ISOEMSs. The basic procedure of the TCS can be described as follows:

1. Every Energy Management System Software Agent (EMSSA) of an FNP n taking part in the TCS auction market sends its bids/offers c_n to the ISOEMS in charge of its proper operation zone[67]. Additionally, predictions of further power infeed and demand of all Inflexible Network Participants (INPs), as defined in Assumption 3.1 below, and operational parameters as e.g. pump and DPR set points, are sent to every ISOEMS by the network operators.

2. All ISOEMSs calculate the optimal control values P_n and Φ_n, in a parallel coordinated form, for all FNPs of the CEPDHN maximizing the social welfare W, while taking into account the operational limits of the CEPDHN. Further, the resulting market prices are determined by the ISOEMSs. Then, the control values and the market prices are sent to the EMSSAs of the FNPs .

This procedure is repeated for several time steps, based on a rolling horizon approach of the NMPC [Grü17]. The bids/offers, predictions, and control values are sent and determined for all time steps of the prediction horizon[68].

[67] Every energy converter is modeled as an FNP in the EPN and DHN. Thus, when an energy converter is located at the border of an operation zone, as shown in Figure 3.2 the EMSSA of this energy converter sends bids/offers to ISOEMSs of both bordering operational zones.

[68] FNPs are incentivized to keep bids/offers for a specific time step constant over the prediction horizon to avoid penalty fees. However, the proposed approach can also handle cases where FNPs change their bids/offers over time.

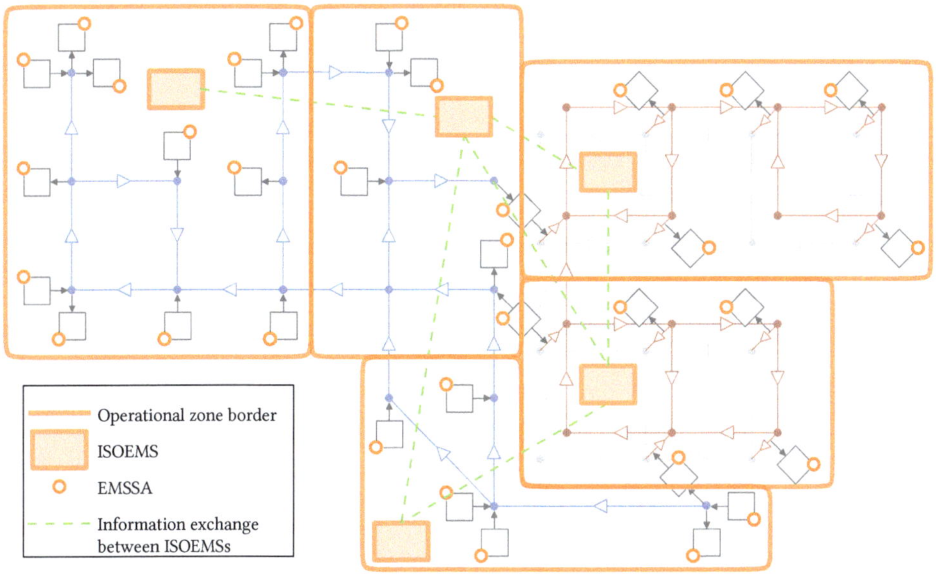

Figure 3.2: Multiple ISOEMSs, each in charge for the operation of its proper operational zone, dispatching the CEPDHN and thereby clearing the intraday TCS auction market in a coordinated form. All elements of the CEPDHN not explained in the given legend are labeled in the legend of Figure 2.1.

The proposed TCS intraday auction market is distinctly different from the traditional wholesale energy markets. This is due to the fact, that the proposed TCS auction is a multi-period combined auction of two or more coupled market platforms associated with the EPN and every DHN, respectively, where these markets are cleared in a coordinated manner and simultaneously.

The energy dispatches in the proposed TCS auction market, operating in 15-minute intervals, change for every time step using a *rolling horizon* technique, in order to optimally adapt to the updated forecasts of supply and demand of inflexible network participants. The bids and offers of the FNPs are assumed to be provided by local EMSSAs, such as existing facility Energy Management Systems (EMSs). These automatically send bids to the ISOEMS for each time step based on the calculated flexibility of their respective facility, such as offices, factories, Battery Energy Storage Systems (BESSs), and/or energy hubs. Observe that these bids and offers are only known to the individual market participant and the TCS auction market operator, thus enhancing privacy as defined in [KW16]. No other information such as indoor temperatures, state of charge of batteries, or electric vehicle data is shared. Note, that privacy is a direct outcome from the information sharing specifications in the proposed TCS. Other approaches, such as direct load control [LLJZ20], which coordinate flexible resources based on monetary incentives, require sharing information not needed in the presented TCS.

The flexibility of consumer facilities can be determined by their EMSSAs by first calculating the predicted minimum power demand for the next hours and the expected usable amount of energy [AB19]. Based on these, two profile bids with minimum and maximum power values can be calculated. The bid prices of the FNPs are determined by their respective EMSSAs using an approach such as [AW14b].

As stated above, the network operators provide predictions to the ISOEMSs, this leads to:

Assumption 3.1. *Sufficiently accurate predictions on the aggregated electric real and reactive power infeed or demand, $\widehat{P}$, and $\widehat{Q}$, of all INPs connected to a specific bus in the EPN, can be provided for every time step on the prediction horizon. Similarly predictions on mass flows $\widehat{\dot{m}}$, differential pressures $\widehat{\Delta p}$ and temperatures $\widehat{T}$ of DHN network parts affiliated to INPs can be provided.*

Projects and literature showing the effective prediction of electric power values at single distribution network buses are given e.g. by [ZSL$^+$17, Sau19]. Besides this practical insight, several approaches can be found within the SOTA, that enable high quality predictions based on measurement data of sensors within the network smart meters at single NPs [MAFH21, WCHK18]. Two additional remarks are necessary here:

Remark 3.1:

For notational simplicity and without loss of generality, within this work it is assumed, that all NPs $n \in \mathbb{S}_n$ are FNPs $n \in \mathbb{S}_n^{\mathrm{fnp}}$, and so $\mathbb{S}_n = \mathbb{S}_n^{\mathrm{fnp}}$. Further, this entails the set of all INPs is empty $\mathbb{S}_n^{\mathrm{inp}} = \varnothing$.

Remark 3.2:

The presented TCS approach[69] is seen as incorporated into the current form of EPN and DHN operation. This means earlier cleared markets and energy trading forms, as e.g. the day-ahead market, futures or Over the Counter (OTC) trading, define the UC . Further, ancillary services as e.g. frequency and voltage regulation in the EPN are present [MBB11]. Similarly in DHNs pressure and expansion maintenance is operating, and the controls of DPRs and pumps are active [NTJK20].

3.2 Market Mechanisms

In this section the market mechanisms used within the context of this work are presented. These are the auction design and market clearing procedure, the allocation objective and the selection of an adequate pricing mechanism. Before these are depicted in detail below, the following assumption is introduced and discussed:

Assumption 3.2. *An intraday heat energy market is implemented in DHNs, if these have a sufficient large amount of MPs in order to limit issues related to market power.*

[69] With the exact definition of a single ISOEMS presented later in Section 3.3.1 or multiple ISOEMSs outlaid in Section 3.3.2.

Note, that this assumption is widely used within literature [LWM+18, ZGX13, LSZW15] due to the following reasons:

1. In DHNs with large amount of small Distributed Energy Resource (DER) infeed, e.g. from CHPs, DHN operators are more and more confronted with uncoordinated heat power infeed, impeding efficient network operation. Here an appropriately designed DHN market would help to coordinate different supply forms (and demand).
2. As pressure raises to save resources and reduce emissions, more and more smart meter systems are installed in DHNs, which would facilitate the implementation of real-time pricing mechanisms [LSZW15].
3. Currently, the DHN intraday market is not highly competitive, due the missing price variations of conventional energy carriers [CWL+18], which are currently used to a large extent for heat generation. Nevertheless, with the need to decarbonize heat supply, this entails a strong electrification of DHN supply, which will bring along larger intraday price variations due to the volatility of RESs [CWL+18]. The coordination of supply and demand is thereby supported by an adequate auction design.
4. Besides the academic discussion about DHN markets, real world implementations of DHN intraday markets are already found in e.g. in Copenhagen [ZWW+21].

Therefore, DHN markets could play an important role in future CEPDHNs, and are thus assumed in operation within this work besides a well known electric energy intraday market.

3.2.1 Auction Design and Market Clearing

When two or more partly coupled energy markets are cleared, here the EPN and multiple DHN markets[70], these markets are either cleared sequentially or simultaneously in the prevailing approaches found in the respective literature [MKP20, DLS+19]. In the case of market based operation approaches for CEPDHNs sequential market clearing approaches typically clear the EPN market first, and then clear the DHN thereafter. Unfortunately, this procedure creates cross-subsidies and limits the feasible allocations of the secondarily cleared market, thus limiting market efficiency. This happens e.g. when using the Alternative Power Method (APM) implemented in Sweden and Finland [DLS+19]. Simultaneous market clearing approaches enable to overcome this downside. However, this is only achieved through information exchange between the Independent Market Operators (IMOs) of the different markets during the market clearing procedure. The TCS presented in this work is built upon a simultaneous market clearing procedure.

Here, a double sided multi-period combined auction is used for all markets within the TCS in order to enable DSM. The market clearing for the EPN and all DHN markets is simultaneously performed for every time step. The time interval Δk between to consecutive time steps is thereby set in the range of current intraday markets, thus within $5\,\text{min} \leq \Delta k \leq 15\,\text{min}$ [Ott03], [KP17]. Therefore, bids and offers for the entire prediction horizon are taken into account, but the market clearing is only carried out for the previous time step. The MPs are

[70] One for each DHN.

incentivized to keep their bids and offers constant, in price and power quantities for a specific time step, during the rolling horizon time step shifts in order to prevent penal fees.

3.2.2 Allocation Objective

The allocation of the multi-period combined auction is derived, based on the welfare W maximization. The latter is defined as the sum of the benefits of all EPN and DHN consumers $n \in \mathbb{S}_n^{\mathrm{d,epn}}$ and $n \in \mathbb{S}_n^{\mathrm{d,dhn}}$, and net costs of all EPN and DHN producers $n \in \mathbb{S}_n^{\mathrm{s,epn}}$ and $n \in \mathbb{S}_n^{\mathrm{s,dhn}}$, for all time steps on the prediction horizon $k \in \mathbb{S}_k$, given as follows:

$$W = \sum_{k \in \mathbb{S}_k} \Bigg[\sum_{n \in \mathbb{S}_n^{\mathrm{d,epn}}} c_{n,k} P_{n,k} + \sum_{n \in \mathbb{S}_n^{\mathrm{s,epn}}} c_{n,k} P_{n,k} + \sum_{n \in \mathbb{S}_n^{\mathrm{d,dhn}}} c_{n,k} \Phi_{n,k} + \sum_{n \in \mathbb{S}_n^{\mathrm{s,dhn}}} c_{n,k} \Phi_{n,k} \Bigg] \tag{3.1}$$

where the bid or offer price per unit is defined by c_n [MCC05], the electric real power injection is given by P and heat power supply is stated as Φ. Note, that consumer powers can only become negative signs in equation (3.1), while producer powers, and all costs have positive values. Storage units may appear either as a consumer (when charging) or as producer (when discharging), and inflexible participants are considered to provide a zero-priced bid/offer, $c_{n,k} = 0$. The ECs of the CEPDHN submit a DHN offer and an EPN bid/offer depending on their type. For example EBs will send EPN bids while CHPs, supplying also electricity, will submit an EPN offer for every time step.

Remark 3.3:
More complex cost functions, e.g. piecewise linear functions resulting from the approximated efficiency curves of generators can be integrated directly into the approach presented here. An FNP can submit multiple bids or offers, one for every segment of the piecewise linear cost function. The allocation based on the optimization of equation (3.1) *ensures that the bids and offers are allocated successively, e.g. first the lowest priced offer of generator/FNP n, then the second lowest priced offer and so on. For notational simplicity and without loss of generality this work is restricted to a single bid or a single offer for consumers and producers.*

3.2.3 Pricing Mechanisms

Four common pricing approaches used to clear energy markets are presented below. These are discriminatory pricing, uniform marginal pricing, zonal pricing and locational marginal pricing. After that, a fifth pricing approach, named as Hybrid Pricing Approach (HPA), developed within the context of this thesis, for the application in CEPDHN market environments is introduced. For that matter, potential benefits and drawbacks occurring by the application of these different pricing methods are presented. A discussion and selection of a pricing approach for the presented TCS is then given after the description of all five pricing methodologies.

Discriminatory Pricing

Following a discriminatory pricing approach the market participants will by paid as bid. The motivation behind this form of pricing is that producers with low energy production costs are supposed to submit low bid prices and will thus be paid low. Following this logic, this would increase the auctioneer's and consumers welfare at the suppliers expense [HL12]. However, the fundamental problem behind discriminatory pricing approaches is the missing incentive compatibility, due to the opportunities for strategic behavior [SBLS04]. This pricing approach is e.g. used in Iran [HL12].

Uniform Marginal Pricing

The Uniform Marginal Pricing Approach (UMPA) relies on insignificance of congestions within the regarded market area [MSC03], this is often referred to as the "copper plate" assumption [CN17, p. 34]. In the tradeoff between market liquidity and price efficiency, UMPA enhances market liquidity [MSC03].

One-shot auctions using the UMPA are proven to be incentive compatible [Vic61, ACP+14]. The Uniform Marginal Price (UMP) is thereby calculated as follows: The IMO aggregates all supply offers into one marginal cost function $f^{\mathrm{s}}(\boldsymbol{x}^{\mathrm{s}})$ by applying the Merit Order (MO). The supplied power of all producers of this allocation is given by $\boldsymbol{x}^{\mathrm{s}}$. Similarly, the demand bids are aggregated vice versa into the marginal cost function $f^{\mathrm{d}}(\boldsymbol{x}^{\mathrm{d}})$, dependent on the power demand $\boldsymbol{x}^{\mathrm{d}}$ of all consumers for the regarded allocation, and both are matched. Assuming an unconstrained lossless network, the intersection of the marginal supply and demand bid cost functions $f^{\mathrm{s}}(\boldsymbol{x}^{\mathrm{s}})$ and $f^{\mathrm{d}}(\boldsymbol{x}^{\mathrm{d}})$ define the UMP c^{ump}[71], and the total amount of power units traded U^{ump}[72]. In this case, incentive compatibility is ensured because no benefit can be achieved by submitting offers higher than the actual cost value for a producer [SRG08a], and likewise in a double-sided auction, this also prevents consumers from submitting bids with costs lower than their real costs. This is due to the fact, that by submitting bids or offers with their marginal costs, MPs will either increase their revenue, or will at least receive their necessary utility. Therefore, in a sealed bid AD with uniform marginal pricing, the participants are incentivized to reveal their marginal costs [Nou95].

In the case of occurring system contingencies, UMPA causes false incentives through inadequate price signals, necessitating command and control interventions in order to maintain secure network operation[8], see [MSC03] for examples on strategic behavior of the MPs. This procedure stands contrary to the desired form of competitive markets, reducing market efficiency due to uplift costs[MSC03]. In Germany, these redispatch costs have risen in the past decades and exceeded 900 million euros in 2017 [Bun18b].

---

[71] Note, that in the unconstrained lossless Security Constrained Economic Dispatch (SCED) case, the UMP is equivalent to the energy component of the LMP presented below.

[72] A definition of the UMP in special cases as an overlap of $f^{\mathrm{s}}(\boldsymbol{x}^{\mathrm{s}})$ and $f^{\mathrm{d}}(\boldsymbol{x}^{\mathrm{d}})$ for multiple points is given in [Zim10].

Zonal Pricing

The Zonal Pricing Approach (ZPA) is used in energy markets in Denmark, Sweden, and Norway [HL12]. In case congestions only occur on very few transmission lines between predefined price zones which can be appointed a priori, zonal pricing is a valuable alternative to UMPA. Advocates of this approach invoke the simplicity of the procedure compared to the much more complex LMPA[73], and that the networks in many cases can be divided naturally into the respective price zones [HL12]. Still, as soon as intra zonal congestions appear, the problems of zonal pricing are equivalent to those of the UMPA, within a zone, again leading to the socialization of uplift costs [MSC03]. Further, real world experience shows, that it is a tough and nontrivial task to predefine adequate price zones. This originates from the fact that congestions are often not rare and do not follow clear spatial patterns [MSC03].

Locational Marginal Pricing

For electric energy market price signals, the LMPA was first presented in [CBS82]. For EPNs the LMPs can be obtained from the dual variables of the power balancing constraints (2.15)[74]. Every single LMP λ^{tot} consists of three price components, the energy price $\lambda^{\mathrm{energy}}$, the congestion price λ^{cong}, and the loss price components λ^{loss}, given as:

$$\lambda_{i,k}^{\mathrm{tot}} = \lambda_{i,k}^{\mathrm{energy}} + \lambda_{i,k}^{\mathrm{cong}} + \lambda_{i,k}^{\mathrm{loss}}, \quad \forall\, i \in \mathbb{S}_i,\ k \in \mathbb{S}_k \tag{3.2}$$

Thus, $\lambda_{i,k}^{\mathrm{tot}}$ expresses the marginal cost to provide the last increment of electric power at a specific bus i in the EPN [MSC03] for a respective time step k. The LMPA inherently overcomes the inefficiencies of the UMPA and the zonal pricing approaches. It comprises uplift costs by directly taking congestions into account, reflecting these in the nodal prices. Hence, the approach inherently calculates optimal price zones for every dispatched solution, which can vary in size from a single node up to the entire network. Multiple studies, listing a slight superiority, by $0.8\,\%$ to $1.5\,\%$ welfare increase, of the LMPA compared to UMPAs are stated in [HL12]. Also studies showing the superiority of the LMPA in contrast to ZPAs are listed in [HL12]. A major advantage of the LMPA is the ability to provide investors with incentives to construct and connect flexibilities at suited buses [MSC03], and thereby reduce congestion and loss costs[75]. However, the LMPA cannot fully prevent strategic behavior and the obtained prices can be dependent on the chosen reference bus [NAB14].

In MES the concept of LMPAs has been applied within the given literature [GA07, DLS+19]. For a DHN, the dual variables of the thermal node equation (2.45)[76] reflect the LMPs. Note, that the thermal node equation needs to be multiplied by the specific heat capacity of water c^{w} before, to obtain properly scaled values for the LMPs. In the case of dynamic network models, as

[73] Which is introduced below.

[74] If the LMPA is implemented within distribution networks, it is often referred to as Distribution Locational Marginal Price (DLMP) [HCC+12]. Thereby, losses should be taken into account, as they are larger as in transmission systems [CWL+18]. The AC power flow equation (2.15) used in this work is thereby often suggested within the literature, see e.g. [OCS04, HCC+12, MC11, CWL+18].

[75] It should be noted, that changing the pricing mechanism to LMPA in the European Energy markets would need high political effort.

[76] Or (2.47) in the case of VMFDs.

often occurring in optimal dispatch problems of DHNs, the LMPs are also dependent on time delay effects [DLS+19]. The time delay can influence all three components of the LMPs mentioned above. These time delay effects still conform with the basic principle of the LMPs, as it charges MPs by the truly arising costs at a specific location in a certain point of time [DLS+19]. However, in DHNs, as FNPs are always connected to supply and return networks, applying the LMPA results in two possible LMPs for every FNP. Furthermore, these two marginal prices for an FNP can differ, due to different constraints applied to the DHN nodes. This happens, e.g. in the context of limiting the temperature ranges of supply and return nodes [MTK+22]. Thus, the assignment of an LMP to each FNP within TCS for (CEP)DHNs remains an open question. In the current form, the LMPA is not suitable for DHNs, as it is not possible to identify a clear buying/selling price for every FNP.

Hybrid Pricing Approach

The description of the HPA is based on the authors previous work [MGR+21, pp. 4-7]. The basic idea behind this approach is to aim for incentive compatibility, and maximize allocation efficiency, while preventing uplift costs due to redispatch. Therefore, in the HPA, uniform marginal pricing is mainly used as it ensures incentive compatibility but also considers network restrictions within the dispatch procedure. However, in the case of network congestion, pure uniform marginal pricing is not incentive compatible, as explained below in this section. Therefore, a fraction of the bids and offers, which exceed U^{umpd}, the amount of power traded at the Uniform Marginal Price obtained from Dispatch (UMPD) c^{umpd}, are then priced via paid-as-bid and pay-as-bid vice versa. The UMPD is defined, as the intersection of the dispatched consumer and producer bid and offer curves, and not by the intersection of the submitted consumer and producer bid and offer functions, in order to take network congestions into account. For simplicity reasons, first lossless energy networks of any domain are assumed.

Remark 3.4:
It is important to distinguish between the UMP c^{ump} obtained from the UMPA introduced in Section 3.2.3 and the UMPD c^{umpd} resulting from the dispatch taking network congestion into account. Therefore the UMPD c^{umpd} is written with an additional "d" for dispatch in the following. Note that, in the case of an uncongested network $c^{\mathrm{ump}} = c^{\mathrm{umpd}}$ always holds.

After the allocation of the power traded between all MPs n, also known as dispatch procedure, the prices paid by all MPs are determined as follows[77]:

The UMPD c^{umpd} is defined as the marginal price at the intersection of the dispatched marginal supply and demand curves $f^{\mathrm{s,dis}}(\boldsymbol{x}^{\mathrm{s,dis}})$ and $f^{\mathrm{d,dis}}(\boldsymbol{x}^{\mathrm{d,dis}})$, see Figure 3.3 (a). All dispatched bids and offers which are to the right of U^{umpd}, the power value at the UMP, also shown in Figure 3.3 (a), are priced via pay-as-bid/paid-as-bid. This is necessary, as in real-world (congested) networks, dispatching the system, while maximizing the welfare, does not always result in a solution, where the total amount of power dispatched is equal to the power value at the UMP U^{umpd}.

[77] The trivial case, in which absolutely no power is traded is left aside here. This would arise if all consumer bids are below all producer offers.

An example of this case is given in Figure 3.3 (a). Here some bids and offers, those bordering the orange shaded area ω^{wf}, need to be priced by via pay-as-bid/paid-as-bid. A detailed description of the reason for this phenomenon is given in Example 3.1 below. Still, this is not always necessary; an example that does not need pay-as bid pricing of some bids is provided in Figure 3.4 (d). Concerning the definition of the UMPD, two special cases are described here:

1. It is possible that there is not only one point of intersection, as bids of consumers and offers of producers may overlap horizontally, see Figure 3.3 (b). Nevertheless, the prices c^{umpd} of all these points are identical. In this case U^{umpd}, the power value at the UMPD is defined as the maximal value of all possible points.
2. Further, it is possible that $f^{\mathrm{s,dis}}(\boldsymbol{x}^{\mathrm{s,dis}})$ and $f^{\mathrm{d,dis}}(\boldsymbol{x}^{\mathrm{d,dis}})$ overlap vertically as depicted in Figure 3.4 (c). In this case, the UMPD is set to the maximum price of the overlapping interval.

As stated above, when optimizing a CEPDHN respecting the operational constraints, a situation may occur where the dispatched power values of $\boldsymbol{x}^{\mathrm{s,dis}}$ and $\boldsymbol{x}^{\mathrm{d,dis}}$ are not equal to the point U^{umpd}, where $f^{\mathrm{s,dis}}(\boldsymbol{x}^{\mathrm{s,dis}})$ and $f^{\mathrm{d,dis}}(\boldsymbol{x}^{\mathrm{d,dis}})$ intersect. This situation arises depending on the network constraints, operational limits, and supply offers and demand bids in the auction. This would not occur, by using the "copper plate assumption", where congestions can not emerge, as in the UMPA approach in Section 3.2.3. However, in reality, network constraints and the operational constraints of the CEPDHN limit the solution set, as shown in the example below:

Example 3.1:

A lossless DC EPN with two buses is shown in Figure 3.5. The respective sources and demands can vary their power demand or infeed between the stated limits at the costs of the offers and bids. The transmission limit is set to 5 p.u.. The result of the social welfare maximization with respect to the transmission capacities is shown in Figure 3.3 (c). Thereby, the transmission capacity between bus 1 and 2 is lower than the amount of power that could be supplied by source S1.

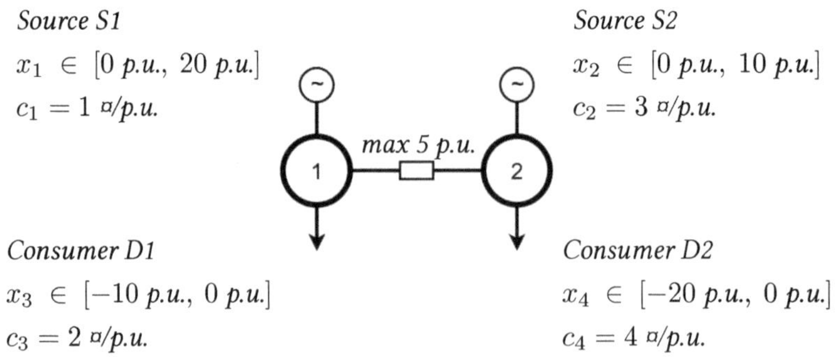

Figure 3.5: An elementary example of a market clearing based on a congested network with two busses connected by a transmission line.

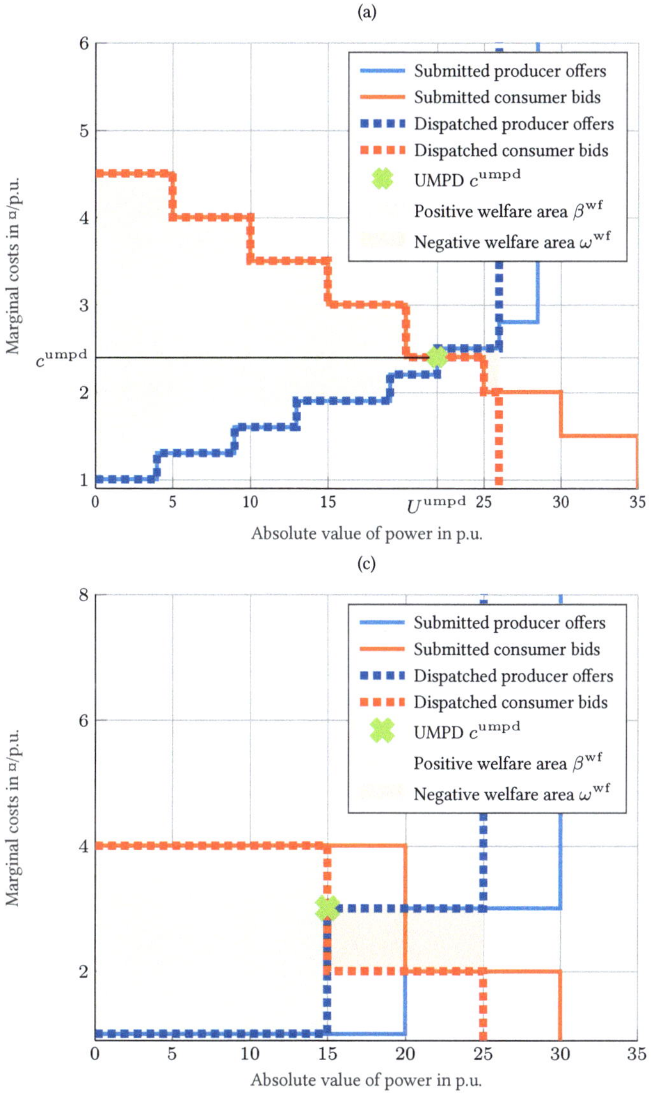

Figure 3.3: Different possible pricing constellations (a) and (c) based on submitted and dispatched offers and bids of producers and consumers sorted by merit order. Further, the resulting UMPDs and welfare areas are shown determined by the optimal dispatch. Note, that the general currency sign ¤ is used here.

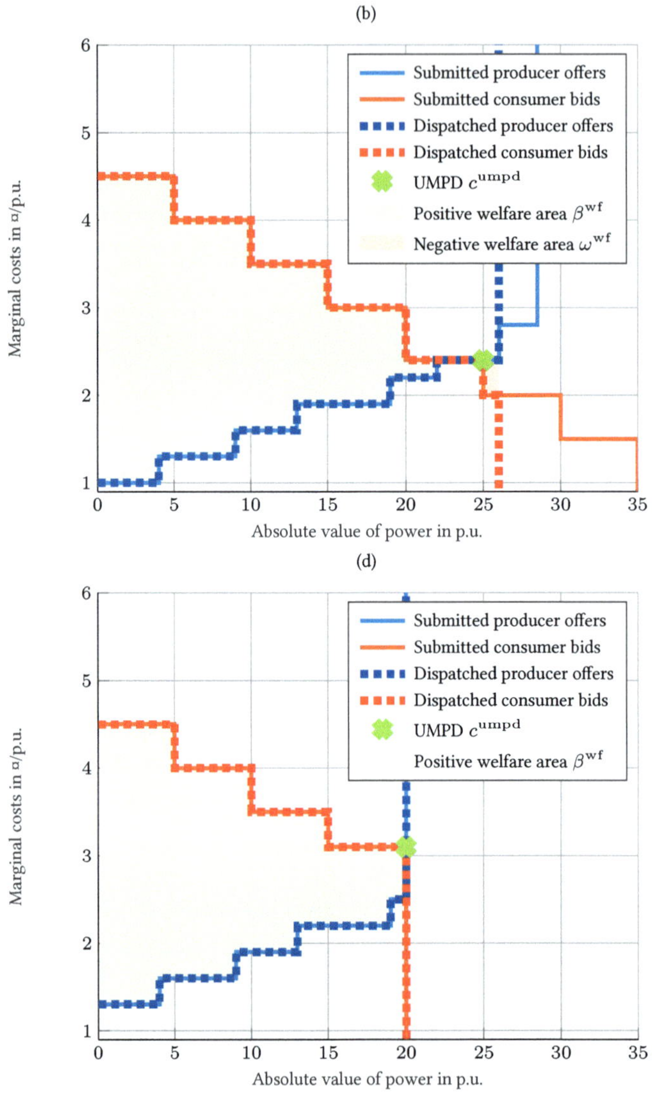

Figure 3.4: Different possible pricing constellations (b) and (d) based on submitted and dispatched offers and bids of producers and consumers sorted by merit order. Further, the resulting UMPDs and welfare areas are shown determined by the optimal dispatch. Note, that the general currency sign ¤ is used here.

In the underlying optimization of the ISOEMS , the total utility of all MPs in the system is maximized. If that allocation is to the right of the intersection of the supply and demand curves as shown in Figure 3.3 (c), the welfare determined by the integral over the orange area ω^{wf} is subtracted from the integral of the green area β^{wf} to receive the final overall system welfare. Since the minimal possible welfare result, based on the optimization is zero, $\beta^{\mathrm{wf}} \geq \omega^{\mathrm{wf}}$, and the optimizer will always try to minimize ω^{wf} while maximizing β^{wf}. The pricing for all MP bids and offers allocated in ω^{wf} is pay-as-bid/paid-as-bid, meaning all consumers pay exactly their bid price and all suppliers earn their respective offer price. The difference between supply and demand prices is paid by the system operator and transferred to all network participants through a grid fee. However, the pay-as-bid/paid-as-bid mechanism encourages strategic behavior. A supplier could gain experience over time and try to place its offer in ω^{wf} and thereby receive higher revenue. Likewise, a consumer could try to set its bid in the same space to pay a lower price. This strategic behavior would go against the incentive compatibility principle implemented with an AD with uniform marginal pricing. Yet, the predictability of RESs is limited and therefore increases the risks resulting from strategic behavior. In a scenario with incomplete or imperfect information, where all participants have neither all nor correct information on all other participants bidding strategies, energy supply or demand and the network topology, the incentive for strategic behavior depends on the size of ω^{wf}. To use the advantage of a UMPA, more precisely, ensuring incentive compatibility to reveal true marginal costs of the market participants, the HPA is efficient for small ω^{wf} compared to β^{wf}. An advantage of the approach compared to a standard UMPA with uplift costs coming from redispatch is, that these costs are directly avoided. This can be understood in the context of the Example 3.1 with Figure 3.5. If source S2 and load D2 would submit offers and bids with the same prices but rising amounts of power, the uplift costs due to redispatch would rise after a standard uniform marginal pricing approach. When applying the HPA presented here, these uplift costs are completely avoided. Instead, a grid fee has to be paid by all network participants to cover the costs of the negative welfare ω^{wf}. As the optimizer inherently minimizes ω^{wf}, this grid fee will stay small no matter how big the offers and bids of source S2 and load D2 would become, in the Example 3.1 stated above. Therefore, this approach will lead to a better overall system efficiency, if the effects of the lost incentive compatibility and the total costs of the grid fee, used to compensate the gap between the bids and offers priced by paid-as-bid/pay-as-bid, are smaller than the gains through the omitted uplift costs. It has to be kept in mind that full incentive compatibility is also currently not given in the uniform marginal pricing-based markets, as bidding parties, for example power plant operators, can be incentivized to bid for redispatch capacities [STRW18]. Finally, following remarks are provided for additional information and clarity:

Remark 3.5:

The consideration of losses and dynamics in network models or ramping constraints is also possible. Similar as network congestions both aspects will change the dispatch result but do not limit the applicability of the pricing procedure of the HPA introduced above. A detailed discussion on the integration of loss bids into the allocation process in order to ensure allocation efficiency is given in [MGR+21]. The main idea is to integrate the losses of the EPN and the DHN as a consumer bid into the allocation objective (3.1). *This ensures allocation efficiency, as also trade for bids and offers with identical marginal costs, is achieved, which would not be the case, if network losses would not be factored during the dispatch procedure. In this matter, the challenge is to determine an adequate marginal bid price* c^{loss} *for the loss bids. The closer the defined value of* c^{loss} *is to the UMPD as defined in this section above, the less undesired side effects, which will be explained in the following, are created. Choosing* c^{loss} *to high will on the one side lead to additional unwanted network losses, causing undesired costs. Note, that expenses created through the loss bids are covered through grid fees paid by all market participants. On the other side, if* c^{loss} *is set to low, the desired market efficiency can not be fully provided. A reasonable choice is to determine* c^{loss} *by the UMP as defined in Section 3.2.3, neglecting network constraints. This enables to use the available a-priori knowledge based on the submitted bids and offers to estimate the new UMPD calculated by taking into account network losses and network constraints. Examples of this procedure are given in the Case Study I in Section 4.1.*

Remark 3.6:
The HPA explained above is performed for every price zone individually after the dispatch, which is carried out for the entire CEPDHN at once in a distributed form. Within the approach, the EPN and every DHN represents a single price zone. This separation of EPN and DHNs price zones is made in order to account for the different bid price levels, as heat is typically a cheaper form of energy as electricity [Kon18, p. 155],[SMA22].

Remark 3.7:
Within the description of the HPA, the pricing is always referred to as a combination of UMPA and pay-as-bid/paid-as-bid. Still, some readers might understand HPA much more as a combination of zonal pricing and pay-as-bid/paid-as-bid. Even though the HPA is used in multiple price zones, the expression zonal pricing will not be used in here for two reasons. First, in contrast to a ZPA the price zones are not determined by a priori knowledge on occurring congestions, but as explained above in Remark 3.6. Further, the CEPDHN is decomposed into multiple operational zones which are controlled by an ISOEMS each. As these zones do not represent the price zones as exemplary depicted in Figure 3.6, the term zonal pricing is not used here to describe the HPA in order to prevent misleading of the readers.

Remark 3.8:
Uniform marginal prices as introduced in Section 3.2.3 can be obtained from the marginal of a balance constraint of power demand, power supply and network losses [DF05], thereby neglecting network constraints. Using this form of determining the UMPDs here would have several drawbacks. First, network constraints would not be taken into account. Second, such a model constraint does not align trivially with the CEPDHN model, introduced in Chapter 2, which includes dynamic effects as for example pipeline storage. Third, using such a balance constraint within the distributed optimization of the ISOEMSs, would necessitate to assign this constraint to one of the resulting subproblems of the operational zones. This would lead to a complicating constraint, see Definition 3.3, which would need variables from all operational zones, for which the UMP is to be calculated. This can create a strong form of coupling among the subproblems which can have negative effects on the convergence properties of distributed optimization approaches as explained in Section 3.3.2.

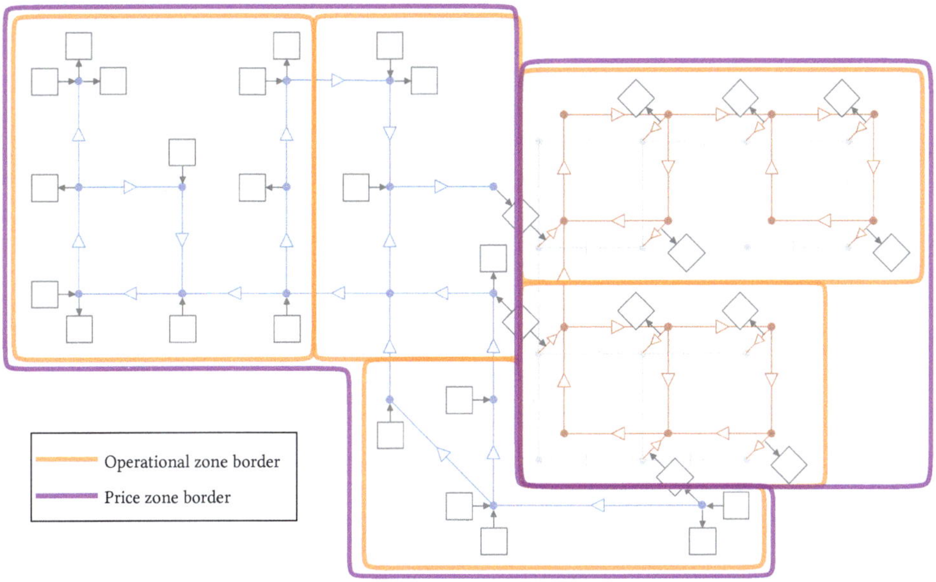

Figure 3.6: Exemplary price zones and operational zones in a CEPDHN from Figure 2.1.

Discussion and Selection

None of the presented approaches is able to prevent incentives for strategic behavior in real world energy markets, especially not in the constrained case, when certain MPs can exploit their market power. However, within pure electric energy markets, the LMPA can be seen as the most efficient pricing approach for market clearing [MSC03]. Still, the application of the LMPA in the TCS for CEPDHNs brings along several problems, that need to be taken into account. First, as mentioned earlier in Section 3.2.3, the LMPA is not suitable for DHNs, as it is not possible to identify a clear buying and selling price for every FNPs. Second, as the TCS is designed as a system of distributed EMSs solving a large optimization problem in parallel, methodical aspects of these distributed optimization approaches need to be considered. For example, the tuning of parameters in the widely used ADMM approach represents a tradeoff between convergence speed and a remaining gap between the dual variables of the central solution to the distributed solution, see Section 3.3.2. This is an undesired feature when using the LMPA in a TCS as convergence speed and accuracy of the LMPs are both of high importance. Third, in order to achieve high technical efficiency, the CEPDHN models become strongly nonlinear. In this context, the utilization of soft constraints can help to consider operational limits within the solution of the optimization problems, also in the case of VMFDs , see Section 3.3.1. However, by using a LMPA , these soft constraints would also directly affect the LMPs creating unwanted deviations. Besides the LMPA, the difficulty of a ZPA, lies in the definition of appropriate zones, for all operation conditions, as outlaid in Section 3.2.3.

Apart from the methodological aspects mentioned above, it has to be recognized, that this work is seen as anchored within the European context, in which a uniform pricing energy market clearing is established[78]. Hence, the implementation of an approach achieving the optimal interaction and coordination of potentially emerging heating markets with the electricity markets would presumably be facilitated by adopting similar pricing mechanisms in the heating markets. Besides, a pure UMPA leads to additional uplift costs, which can entailed high costs, as discussed in Section 3.2.3. This is why, this work is based on the HPA presented in the previous Section 3.2.3. Using this approach, incentive compatibility is achieved under the conditions mentioned before, and the different price levels in the EPN and the DHNs are adequately represented. Summarizing the requirements stemming from the auction design, the market clearing procedure, and the HPA, leads to the following statement:

Assumption 3.3. *The EPN and all DHN markets are cleared in a joint, coordinated procedure simultaneously, while taking into account physical network constraints during welfare maximization.*

In doing so, additional uplift costs due to redispatch are prevented.

3.3 Control Mechanisms

The control mechanisms regarded and chosen within the TCS design process are outlaid in the following sections. The basic problem structure of the ISOEMSs and the used class of control approach is presented first introducing the necessary terminology. Then Section 3.3.1 explains how a central ISOEMS controller can be designed in detail. This is followed by the design of the ISOEMSs using a distributed control approach in Section 3.3.2.

In this work a rolling horizon control mechanism is used as it enables to maximize a general predefined objective, taking system constraints into account, including measurement data and predicted INP power demand and supply as well as operational parameters as pump and DPR operational set points. Within the TCS for CEPDHNs presented here, the objective is to maximize the welfare W, resembling the market allocation objective defined in equation (3.1). Whereas, the system constraints are determined by the network models, given in Chapter 2, and the operational limits, stated in the following Section 3.3.1.

[78] Within the different price zones.

Within the context of control engineering, the ISOEMSs within the TCS represent a *nonlinear distributed economic time-variant hybrid discrete model predictive control approach* for specific flow conditions within the DHNs. The terminology is explained as follows:

- *Nonlinear Model Predictive Control (NMPC)* : Given a nonlinear optimal control problem, following [DFH09, p. 393]:

$$\min_{\boldsymbol{x}^{\mathrm{st}},\boldsymbol{z}^{\mathrm{st}},\boldsymbol{u}} \sum_{k\in\mathbb{S}_k} q^{\mathrm{obj}}(\boldsymbol{x}_k^{\mathrm{st}}, \boldsymbol{z}_k^{\mathrm{st}}, \boldsymbol{u}_k) \tag{3.3a}$$

subject to

$$\boldsymbol{x}_1^{\mathrm{st}} - \overline{\boldsymbol{x}}^{\mathrm{st},0} = \mathbf{0}, \tag{3.3b}$$

$$\boldsymbol{x}_{k+1}^{\mathrm{st}} - \boldsymbol{f}_k^{\mathrm{sys}}(\boldsymbol{x}_k^{\mathrm{st}}, \boldsymbol{z}_k^{\mathrm{st}}, \boldsymbol{u}_k) = \mathbf{0}, \quad \forall\, k \in \mathbb{S}_k, k < |\mathbb{S}_k| \tag{3.3c}$$

$$\boldsymbol{h}_k(\boldsymbol{x}_k^{\mathrm{st}}, \boldsymbol{z}_k^{\mathrm{st}}, \boldsymbol{u}_k) = \mathbf{0}, \quad \forall\, k \in \mathbb{S}_k \tag{3.3d}$$

$$\boldsymbol{g}_k(\boldsymbol{x}_k^{\mathrm{st}}, \boldsymbol{z}_k^{\mathrm{st}}, \boldsymbol{u}_k) \leq \mathbf{0}, \quad \forall\, k \in \mathbb{S}_k \tag{3.3e}$$

  with the system state space variables of all time steps $\boldsymbol{x}^{\mathrm{st}} = [\boldsymbol{x}_1^{\mathrm{st}}, ..., \boldsymbol{x}_k^{\mathrm{st}}, ..., \boldsymbol{x}_{\mathrm{n}^{\mathrm{ck}}}^{\mathrm{st}}]$, the algebraic state variables $\boldsymbol{z}^{\mathrm{st}} = [\boldsymbol{z}_1^{\mathrm{st}}, ..., \boldsymbol{z}_k^{\mathrm{st}}, ..., \boldsymbol{z}_{\mathrm{n}^{\mathrm{ck}}}^{\mathrm{st}}]$ and the control values given by $\boldsymbol{u} = [\boldsymbol{u}_1, ..., \boldsymbol{u}_k, ..., \boldsymbol{u}_{\mathrm{n}}^{\mathrm{ck}}]$ with the cardinality of all time steps on the prediction horizon given by $\mathrm{n}^{\mathrm{ck}} = |\mathbb{S}_k|$. Further, the initial system state is defined by $\overline{\boldsymbol{x}}^{\mathrm{st},0}$. The nonlinear discrete dynamic system model is given in equation (3.3c). Together, the equality constraints described with $\boldsymbol{h}$ in equation (3.3d), and the nonlinear discrete dynamic model (3.3c) represent forms a Differential Algebraic Equation (DAE) form [DFH09, p. 392f]. Additionally, the system boundaries are aggregated in $\boldsymbol{g}$. Solving the discrete dynamic nonlinear optimization problem (3.3) for every time step, in order to find the optimal control values $\boldsymbol{u}^*$ minimizing the objective q^{obj}, taking into account latest system state measurements $\overline{\boldsymbol{x}}^{\mathrm{st},0}$ is referred to as MPC [BBM17]. Using the *nonlinear* system model (3.3c) is usually referred to as NMPC [Grü17]. The term *discrete* stems from the discrete system description.

- *Economic*: NMPC applications, where the objective q^{obj} "does not penalize the distance to a predefined equilibrium" [Grü17, p. 221], but for example the operation costs of systems, are referred to as *economic* NMPC.

- *Distributed*: When multiple NMPCs, each solving a subproblem of (3.3) originating from a decomposition of the central problem (3.3) this is described as *distributed* NMPC. Thereby, the individual NMPCs share information during the numeric solution of their respective sub problem. The individual sub problems may thereby "be linked by dynamics, cost functions or constraints" [Grü17, p. 259].

- *Time-variant*: In these systems, the system model $\boldsymbol{f}^{\mathrm{sys}}$ is time dependent, thus written as $\boldsymbol{f}_k^{\mathrm{sys}}$.

- *Hybrid*: Here, additionally an event based subsystem is used, which detects the current event ev taking place. The system model $\boldsymbol{f}_{ev}^{\mathrm{sys}}$ is then dependent on the current event [EFS03, p. 67ff].

For the ISOEMSs in the TCS used for the control of the CEPDHN the following assignment of variables to $\boldsymbol{x}^{\text{st}}$, $\boldsymbol{u}$, and $\boldsymbol{z}^{\text{st}}$ can be made:

- System states $\boldsymbol{x}^{\text{st}}$: Temperature of water masses in pipelines T^{wm}, originating from the entering node temperatures, and SOCs of large TESSs E, see Remark 3.10.
- Control values $\boldsymbol{u}$: Power infeed and demand of FNPs in EPN P, and Q, and the DHN Φ, as well as the operational set points of valves and pumps, K, and Δp^{pump}.
- Algebraic state variables $\boldsymbol{z}^{\text{st}}$: Defined by the rest of all variables.

In control theory, many control approaches, as NMPC are based on a (nonlinear) state space model of the controlled system. In the case of the CEPDHN controlled by the ISOEMS , obtaining such a regular state space model is not possible, as explained in the following Remark 3.9.

Remark 3.9:

Assuming constant mass flows and thus also constant flow directions, a pipeline can be modeled as a (damped) dead-time element model, which can be formulated as a discrete dynamic state space system model as in (3.3c). *However, this will only lead to accurate results, if the dead-time t^{dead} can be expressed through multiple time step intervals $t^{\text{dead}} = n\Delta k$, with $n \in \mathbb{N}^{>0}$, see [Lun14b, p. 113f], [Lun14c, p. 441]. Therefore, this would necessitate a very small time step interval Δk in order to achieve the demanded accuracy, which would entail a large number of optimization variables. Besides, the assumption of constant flow conditions is not valid if highly efficient DHN operation is to be achieved, see Section 1.2.5, as aimed for in this work. Note, that varying flow velocities[79] lead to changing dead-times, and therefore to a variation of the system model. Thus, this results in a hybrid system with a hybrid time-invariant state space model, where the events are a change of the mass flow within a pipeline.*

Besides the pipelines, large TESSs represent the second CEPDHN components which are represented within equation (3.3c).

Remark 3.10:

TESSs are distinguished into two groups. The first, are large storages which are owned and operated by the network operator. As these are directly connected to the DHN [80], meaning that no heat exchanger is used to separate the water circulation of the DHN from the storage, they are seen as part of the modeled primary network[81] and thus their thermal and hydraulic interaction with the other network components and their SOC is explicitly taken into account by the model, see Section 2.3.3. The losses of these water tanks l are time dependent[82], see equation (2.65), *and thereby an example for the time variance of the system model* (3.3c). *The second group of TESS are owned and operated by FNPs and indirectly connected to the DHN through heat exchangers. For*

79 Assuming constant flow direction.

80 Assuming sensible hot water storages as defined in Section 2.3.3.

81 Water circuit of the main pipelines of the DHN [NTJK20]. All indirectly connected producers and consumers have their own secondary networks behind the heat exchanger. In large DHNs secondary networks can also be used to distribute heat at lower temperatures as in the primary network to local consumers, then the local network participants have their own tertiary networks.

82 As they are dependent on the mean temperature of the water within the storage system $T^{\text{stor,mean}}$ which is time dependent.

these storages, the SOC is not known by the ISOEMS, and thus equation (2.63)[83] *are not taken into account. Thus, also no state variables are assigned to these storages. Thus, they are regarded by the approach as regular consumers or producers.*

Also, note, that a large TESS is not necessarily found in every Zone assigned to an ISOEMS. Further, also other constraints, as ramping constraints introduced later on in Section 3.3.1 bring along dynamics, which can not be incorporated in (3.3), as (3.3e) is only dependent on a single time step.

Due to the aspects mentioned above, the overall control approach implemented in the TCS is referred to as a *distributed rolling horizon approach*, which is explained in detail in the following subsections. For that matter, first the design of a central ISOEMS is presented, which is then extended to the distributed case.

3.3.1 Centralized ISOEMS

Within this Subsection the rolling horizon approach, used in case of a single ISOEMS controlling the entire CEPDHN is designed. The optimization problem solved within every time step k by the ISOEMS can be generalized as the following nonlinear problem:

$$\min_{\boldsymbol{x}} f(\boldsymbol{x}) \tag{3.4a}$$

subject to

$$\boldsymbol{h}(\boldsymbol{x}) = \boldsymbol{0}, \tag{3.4b}$$

$$\boldsymbol{g}(\boldsymbol{x}) \leq \boldsymbol{0} \tag{3.4c}$$

Therein, $\boldsymbol{x}$ represents the vector of all optimization variables. Note, that a classification of the variables in $\boldsymbol{x}$ into system states $\boldsymbol{x}^{\mathrm{st}}$, algebraic state variables $\boldsymbol{z}^{\mathrm{st}}$, and control values $\boldsymbol{u}$ is left aside from here on, due to the aspects mentioned in Remarks 3.9, and 3.10. Also, this helps to facilitate the notation, especially considering the distributed optimization approaches presented in Section 3.3.2. The exact objective function f and the operational constraints $\boldsymbol{g}$, used by the ISOEMS, are depicted below in the *Objective Function* part in Section 3.3.1 and the *Operational Constraints* part in Section 3.3.1. The equality constraints combined in $\boldsymbol{h}$ represent the CEPDHN model summarized in (2.105) and introduced in Chapter 2.

The overall operation procedure of the ISOEMS within the TCS is given in the Nassi-Shneidermann diagram in Figure 3.7.

Two significantly important design parameters are the time step length Δk and the prediction horizon, understood as the amount of future time steps taken into account $n^{\mathrm{p}} = n^{\mathrm{ck}} = |\mathbb{S}_k|$. It is a commonplace that Δk is to be chosen as small as possible, to achieve high model accuracy, while n^{p} should be chosen as large as possible, in order to enable foresighted control actions. However, both of these incentives tend to enlarge the amount/vector of the optimization variables $\boldsymbol{x}$ of (3.4). Thus, a tradeoff between small temporal discretization, length of the prediction horizon, and the resulting computation time needed to solve (3.4) needs to be

[83] And consequently also equations (2.64) and (2.65).

Initialization of the ISOEMS

For every time step k

1. Every EMSSA of an FNP n taking part in the TCS auction market sends its bids/offers c_n to the ISOEMS. Additionally, predictions of further power infeed and demand of all INPs, as defined in Assumption 3.1 above, and operational parameters as e.g. pump set points, are sent to the ISOEMS by the network operator.

2. The ISOEMS calculates the optimal control values for all FNPs, and hydraulic devices by solving the NLP problem (3.4) and sends these to the respective devices for the upcoming time step.

3. The ISOEMS calculates the respective price for every bid and every offer, based on the HPA, mainly using uniform marginal pricing, see Section 3.2.3. These prices are then sent to the FNPs.

4. All provided measurements from sensors in the CEPDHN are used to update variables in $\boldsymbol{x}$, and parameters in the system model $\boldsymbol{h}$. Typical sensor data available in CEPDHNs comprises measurements of mass flow, pressure potentials, temperatures, voltages, and currents.

5. The rolling horizon time shift is performed for all variables in $\boldsymbol{x}$, and all parameters in $\boldsymbol{h}$. The basic rolling horizon procedure for all variables in $\boldsymbol{x} = [\boldsymbol{x}_{k_1}, ..., \boldsymbol{x}_k, ..., \boldsymbol{x}_{k_{n^{\mathrm{nk}}}}]$, replaces $\boldsymbol{x}_k = \boldsymbol{x}_{k+1}$, and sets the newly last time step equal to the previous time step $\boldsymbol{x}_{k_{n^{\mathrm{nk}}}} = \boldsymbol{x}_{k_{n^{\mathrm{nk}}-1}}$ with $n^{\mathrm{nk}} = |\mathbb{S}_k|$. In contrast to this procedure, the rolling horizon time step shift of the thermal pipeline parameters $Z_{e,k,\zeta,\sigma,\tilde{k},\tilde{\sigma}}$, $Z^{\mathrm{P}}_{e,k,\zeta,\sigma,k^{\mathrm{P}},\tilde{\sigma}}$, $\widehat{\dot{m}}_{e,k,\zeta,\sigma}$ and $\kappa_{e,k,\zeta,\sigma}$, need a special procedure, described in Section 2.3.3.

6. Based on the previous parameter calculation, the auxiliary parameters of the thermal pipeline model $\gamma_{e,k,\sigma}$, $\varepsilon_{e,k,\sigma}$, $R^{\mathrm{wm}}_{e,k,\sigma}$, $S^{\mathrm{wm}}_{e,k,\sigma}$, $w_{e,k,\zeta,\sigma}$, and $t_{e,k,\sigma}$ can be updated, as stated towards the end of Section 2.3.3.

Figure 3.7: Nassi-Shneidermann diagram of ISOEMS operation procedure including the rolling horizon approach.

performed [ZWW+21]. Through a reasonable choice, this enables to calculate $\boldsymbol{x}^*$ the optimal solution of (3.4) within Δk during operation. A reasonable choice of the time step length Δk is within $5\,\text{min} \leq \Delta k \leq 15\,\text{min}$. These ranges are typically applied in Economic Dispatch (ED) [Arn11, p. 26] and are thus also a valid choice here. Further, as these time step ranges are also applied in intraday markets, as introduced already in Section 3.2.1, it is reasonable to use the same time step interval within the market and control mechanisms, thereby enabling technical efficiency and economic optimality simultaneously. Possible choices for prediction horizons will be discussed within the results in Chapter 4 in detail.

Objective Function

Typical flow velocity limits in DHNs are given in the range of $0.8\,\text{m/s}$ to $3\,\text{m/s}$ [Glü85b, p. 29]. The lower limits are applied to prevent deposits and the overheating of pumps [NTJK20, p. 124]. In the case of VMFDs it is not possible to implement box constraints, as in the following Section 3.3.1 for the mass flows, which would bound the flows away from zero $|\dot{m}| > \Delta\varepsilon$, with $\Delta\varepsilon > 0$. Note, that this would prevent possible flow direction changes, defining these by the initial flow directions.

This can be faced, by introducing soft constraints into the objective function f of the ISOEMS in problem (3.4), which incentivizes the solver to find a solution $\boldsymbol{x}^*$ in which the mass flows are slightly bounded away from zero $|\dot{m}| > \Delta\varepsilon$, with $\Delta\varepsilon > 0$. Therefore, the objective is put together by the sum of the following two parts:

$$f(\boldsymbol{x}) = f^{\text{eco}}(\boldsymbol{x}) + f^{\text{soft}}(\boldsymbol{x}) \tag{3.5}$$

Therein the economic objective f^{eco}, is defined as the allocation objective, the negative signed welfare W stated in equation (3.1), resulting in $f^{\text{eco}} = -W$. Further, the soft constraints are given as f^{soft}. These soft constraints are implemented by using continuous differentiable approximation of of the rectangular function:

$$f^{\text{soft}}(\boldsymbol{x}) = \sum_{e \in \mathbb{S}_{\text{e}}^{\text{vmfd}}} \sum_{k \in \mathbb{S}_k} f^{\text{rect}}(\dot{m}_{e,k}) \tag{3.6}$$

with

$$f^{\text{rect}}(\dot{m}_{e,k}) = w_{e,k}^{\text{rect,h}} \left[1 + \text{sgn}_{\Delta\varepsilon}\left(\mathrm{w}_{\mathrm{e,k}}^{\text{rect,w}} - (\dot{\mathrm{m}}_{\mathrm{e,k}})^2 \right) \right] \tag{3.7}$$

The weighting factors $w_{e,k}^{\text{rect}}$ define the form of the rectangular function approximation. In doing so, the width of f^{rect} is enlarged by $w_{e,k}^{\text{rect,w}}$, while the height is set by $w_{e,k}^{\text{rect,h}}$, as shown in Figure 3.8. The used continuously differentiable approximation of the signum function in equation (3.7) is given by:

$$\text{sgn}_{\Delta\varepsilon}(x) = \frac{x}{|x|_{\Delta\varepsilon}} \tag{3.8}$$

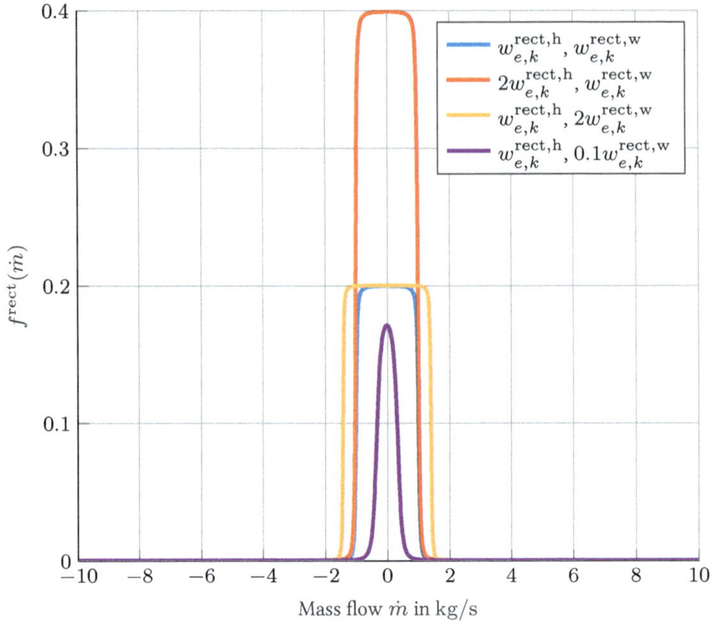

Figure 3.8: Approximation of the rectangular function, given weighting factors $w^{\text{rect,w}}_{e,k} = 1$ and the height $w^{\text{rect,h}}_{e,k} = 0.1$ of f^{rect}.

while the continuous differentiable approximation of the absolute value function is defined as in equation (2.20).

Remark 3.11:

Besides the operational benefits, the presented soft constraints f^{soft} also provide a second salient feature. Reports on convergence issues of optimization problems, stemming from the consideration of VMFDs are found in [ZWW+21] and [Trö99]. Also within the DHN model presented in Chapter 2, in some cases, convergence issues were seen, especially when mass flows were near zero $\dot{m} \approx 0$. In this context, using the soft constraints f^{soft} to slightly bound $\dot{m}$ away from zero enhances the convergence properties.

Operational Constraints

The specification of operational constraints is given below for the EPN, and the DHN respectively. These boundaries not only ensure secure and preserving network operation but also improve the NLP solver performance [Cor20, p. 319]. The variable bounds of the EPN, and the FNPs connected to it are:

$$V_i^{\min} \le V_{i,k} \le V_i^{\max}, \qquad \forall\, i \in \mathbb{S}_i^{\mathrm{epn}}, k \in \mathbb{S}_k \tag{3.9}$$

$$P_{n,k}^{\min} \le P_{n,k} \le P_{n,k}^{\max}, \qquad \forall\, n \in \mathbb{S}_n^{\mathrm{epn}}, k \in \mathbb{S}_k \tag{3.10}$$

$$Q_{n,k}^{\min} \le Q_{n,k} \le Q_{n,k}^{\max}, \qquad \forall\, n \in \mathbb{S}_n^{\mathrm{epn}}, k \in \mathbb{S}_k \tag{3.11}$$

Voltage amplitudes V at buses i are kept within minimum and maximum values, usually defined in the per unit system. The limits on active and reactive power are defined by relevant distribution system transformer capacities, in the EPN , and represent the bid and offer limits set by the EMSSAs of the FNPs, and may vary over time. Note, that feeder limits can also be readily included in the proposed EPN model.

The inequalities used to represent the operational DHN, and the connected FNPs, limits are:

$$p_i^{\min} \le p_{i,k} \le \Delta p_i^{\max}, \qquad \forall\, i \in \mathbb{S}_i^{\mathrm{dhn}},\ k \in \mathbb{S}_k \tag{3.12}$$

$$\Delta p_e^{\min} \le \Delta p_{e,k} \le \Delta p_e^{\max}, \qquad \forall\, e \in \mathbb{S}_e^{\mathrm{dhn}},\ k \in \mathbb{S}_k \tag{3.13}$$

$$\dot{m}_e^{\min} \le \dot{m}_{e,k} \le \dot{m}_e^{\max}, \qquad \forall\, e \in \mathbb{S}_e^{\mathrm{dhn}},\ k \in \mathbb{S}_k \tag{3.14}$$

$$K_e^{\min} \le K_{e,k} \le K_e^{\max}, \qquad \forall\, e \in \mathbb{S}_e^{\mathrm{vlv}},\ k \in \mathbb{S}_k \tag{3.15}$$

$$T_{i,k}^{\min} \le T_{i,k} \le T_{i,k}^{\max}, \qquad \forall\, i \in \mathbb{S}_i^{\mathrm{dhn}}, k \in \mathbb{S}_k \tag{3.16}$$

$$E_n^{\min} \le E_{n,k} \le E_n^{\max}, \qquad \forall\, n \in \mathbb{S}_n^{\mathrm{stor}}, k \in \mathbb{S}_k \tag{3.17}$$

$$T^{\min} \le T_{e,k}^{\mathrm{out}} \le T^{\max}, \qquad \forall\, e \in \mathbb{S}_e^{\mathrm{exch}}, k \in \mathbb{S}_k \tag{3.18}$$

$$\Phi_{n,k}^{\min} \le \Phi_{n,k} \le \Phi_{n,k}^{\max}, \qquad \forall\, n \in \mathbb{S}_n^{\mathrm{dhn}}, k \in \mathbb{S}_k \tag{3.19}$$

Pressure potentials p need to be maintained within respective bounds at all nodes i, since the minimum limits $p_i^{\min}$ prevent evaporation and corrosion and maximum limits $p_i^{\max}$ prevent damage of the network components. Similarly, this security aspect applies to the differential pressures over DHN edges Δp, where the lower bounds also guarantee satisfactory mass flow to meet the power demand of consumers. As already mentioned in Section 3.3.1 the lower bounds on mass flows $\dot{m}_e^{\min}$ prevent the accumulation of deposits and the overheating of pumps, while the upper bounds $\dot{m}_e^{\max}$ prohibit high pressure losses and thereby lower the power necessary for operation of the pumping devices. Note, that for edges e with VMFDs the lower bound is defined as $\dot{m}_e^{\min} = -\dot{m}_e^{\max}$, while soft constraints bound the mass flows away from zero, see Remark 3.11. For edges with fixed flow directions, for example due to a check valve in the respective *string of edges*[84] $\dot{m}_e^{\min} > 0$, is used to define the flow direction within the operation NLP problem (3.4). The valve flow factor limits are used to represent the two extreme values of operation; a closed valve for $K_e^{\min}$ and an open valve for $K_e^{\max}$. The node

[84] A sequence of edges, where every node only connects two edges of the regarded sequence.

Table 3.1: Typical boundaries values of most relevant operational variables in CEPDHNs.

Variable	**Minimum**	**Maximum**
Voltage amplitude V_i	0.85 p.u. to 0.9 p.u.	1.1 p.u. [MCC+13]
Nodal pressure p_i	$> p^{\text{steam}}(T_i) + 1.5\,\text{bar}$[85]	$< 25\,\text{bar}$[86] [ZWW+21]
Mass flow $\dot{m}$	$\dot{m}^{\min} = v^{\min} \rho^{\text{w}} A^{\text{cross}}$, $v^{\min} = -3 \ldots 0.8\,\text{m/s}$ [87]	$\dot{m}^{\max} = v^{\max} \rho^{\text{w}} A^{\text{cross}}$, $v^{\max} = 3\,\text{m/s}$ [Glü85b, p. 29],[SZP+09]
Node temperature T_i	$\max(T^{\text{freeze}}, T^{\text{a}})$[88]	120 °C to 140 °C [ZWW+21, NTJK20]

temperatures T_i are limited in the supply network to guarantee satisfactory heat power provision to consumers and prevent damage of network components; when applied, limits on the return network temperatures enable a high operation efficiency with low losses. The possible SOC of a TESS E is limited by the maximal admissible water temperature and the volume of the TESS. Minimal outlet temperatures of FNPs $T_{e,k}^{\text{out}}$ can be set, if the prediction horizon is insufficient to exhibit, that lower values of $T_{e,k}^{\text{out}}$ will lead to nodal temperatures at the last critical consumer going below the temperature limits $T_{i,k}^{\min}$, necessary for sufficient heat supply. Limits on heat power infeed or demand $\Phi_{n,k}^{\min}$ and $\Phi_{n,k}^{\max}$ originate from the bids and offers of the FNPs.

Remark 3.12:

In cases, where variables of $\boldsymbol{x}$ appear in the demoninator of a specific model equation, the operational limits should be bounded away slightly from zero by setting $x^{\min} = \Delta\varepsilon$, with sufficiently small $\Delta\varepsilon > 0$. This applies e.g. for the minimum valve flow factor $K_e^{\min}$, see equation (2.31).

In general, it is not possible to give one set of parameters applicable to every CEPDHN for all operation conditions, due to large differences in design and operation form. Still, for the most important variable limits of the operational bounds mentioned above, parameter ranges are stated in Table 3.1. Ramping constraints can be applied if needed to represent limits in electric power injection changes of FNPs[89]:

$$P_{n,k-1} - \Delta P_n^{\text{ramp}} \leq P_{n,k} \leq P_{n,k-1} + \Delta P_n^{\text{ramp}}, \qquad \forall\, n \in \mathbb{S}_n^{\text{epn}}, k \in \mathbb{S}_k \tag{3.20}$$

$$Q_{n,k-1} - \Delta Q_n^{\text{ramp}} \leq Q_{n,k} \leq Q_{n,k-1} + \Delta Q_n^{\text{ramp}}, \qquad \forall\, n \in \mathbb{S}_n^{\text{epn}}, k \in \mathbb{S}_k \tag{3.21}$$

The FNP ramping limits are given by ΔP_n^{ramp} for real and ΔQ_n^{ramp} for reactive power.

85 "This means that at the highest point of the network the pressure of the district heating water must be at least 0.5 bar higher than the steam pressure at the maximum network temperature [...]. An additional 1 bar is recommended for any unexpected events."[NTJK20, p. 52]

86 "the pressure at maximum load at a specified point and at maximum network temperature"[NTJK20, p. 52].

87 Typical pipeline diameters are in the range of 20 mm to 1000 mm [NTJK20, p. 64]. Also see Remark 3.11

88 The maximum value of a value below the freezing temperature T^{freeze} and the ambient temperature T^{a}, are possible from a technical stand point. Often return network temperatures do not fall below 35 °C [ZWW+21].

89 For k_1, the previous time step $k_1 - 1$ is equal to the first past time step k_1^{P}.

Similarly ramp limits are applicable for the DHN operation, as[89]:

$$\Delta p_{e,k-1} - \Delta p_e^{\text{ramp}} \leq \Delta p_{e,k} \leq \Delta p_{e,k-1} + \Delta p_e^{\text{ramp}}, \qquad \forall\, e \in \mathbb{S}_e^{\text{dhn}},\ k \in \mathbb{S}_k \tag{3.22}$$

$$\dot{m}_{e,k-1} - \Delta \dot{m}_e^{\text{ramp}} \leq \dot{m}_{e,k} \leq \dot{m}_{e,k-1} + \Delta \dot{m}_e^{\text{ramp}}, \qquad \forall\, e \in \mathbb{S}_e^{\text{dhn}},\ k \in \mathbb{S}_k \tag{3.23}$$

$$T_{i,k-1} - \Delta T_i^{\text{ramp,node}} \leq T_{i,k} \leq T_{i,k-1} + \Delta T_i^{\text{node}}, \qquad \forall\, i \in \mathbb{S}_i^{\text{dhn}}, k \in \mathbb{S}_k \tag{3.24}$$

$$T_{e,k-1}^{\text{out}} - \Delta T_e^{\text{ramp,out}} \leq T_{e,k}^{\text{out}} \leq T_{e,k-1}^{\text{out}} + \Delta T_e^{\text{ramp,out}}, \qquad \forall\, e \in \mathbb{S}_e^{\text{exch}}, k \in \mathbb{S}_k \tag{3.25}$$

$$\Phi_{n,k-1} - \Delta \Phi_n^{\text{ramp}} \leq \Phi_{n,k} \leq \Phi_{n,k-1} + \Delta \Phi_n^{\text{ramp}}, \qquad \forall\, n \in \mathbb{S}_n^{\text{dhn}}, k \in \mathbb{S}_k \tag{3.26}$$

Ramping limits on differential pressure Δp_e^{ramp} or mass flows $\Delta \dot{m}_e^{\text{ramp}}$ are used to limit set point changes of pumps which help to prevent material fatigue. Similarly, the limits on temperature differences between time steps $\Delta T_i^{\text{ramp,node}}$ helps to reduce aging of pipeline insulation, see [MMW+20] or Section 2.99. This can also be applied to the FNPs, where a restriction of the variation of the output temperature $\Delta T_n^{\text{ramp,out}}$, and the heat power $\Delta \Phi_n^{\text{ramp}}$, can be used to prevent fast aging of heat exchangers.

3.3.2 Multiple Distributed ISOEMSs

There are two main reasons motivating the design of a *distributed* rolling horizon approach, with multiple distributed ISOEMSs, solving the operation NLP problem (3.4), in a coordinated form. The first reason is to reduce the computational burden of the potentially large scale NLP problem solved for every time step by parallelization [CNP02]. The second is the ability to reduce the amount of information sharing between different stakeholders, network operation systems and NPs. Note, that as stated in [LGZ+19]: "for the aim of privacy protection, passing some sensitive information such as the system configurations, network data, building parameters, and user preferences to the other party is not acceptable." Thus, a TCS designed for a CEPDHN reaching over the areas of multiple network operators, could be possibly implemented more simply if the operational zones of the ISOEMSs were aligned with the network boundaries of the different network operators. Apart from Germany, in many regions Distribution System Operators (DSOs) do not operate EPNs and DHNs but these are operated by different institutions [LWM+18].

Remark 3.13:
Optimal dispatch problems with dynamic models, ranging over multiple time steps, can be decomposed into subproblems by two different forms. The first form divides the underlying network models into multiple operational zones, thereby applying a spatial decomposition [MDS+17]. The second separates the time steps into multiple sets with consecutive time steps [RMIMD13]. In this work, only spatial decompositions are taken into account, as in most sources on distributed optimal dispatch of energy systems, see e.g. [MDS+17]. This is done as it is expected that the dynamics of the thermal pipeline model presented in Section 2.3.3, the TESSs model provided in Section 2.3.3, and the ramping constraints stated in Section 3.3.1 necessitate more information exchange and create more complicating constraints[90] in the case of a temporal decomposition, as the network models in case of spatial decomposition over tie lines and pipelines. Note, that alone the pipeline

[90] A constraint that is dependant on variables from a different subproblem (here operational zone), see Definition 3.3.

model would coupled multiple time steps, for every time step k and every pipeline in the regarded DHN, through the mapping of the node temperatures to the positions in the respective pipelines and the outlet temperatures, see (2.70) *which incorporates* (2.66). *Additionally, a temporal decomposition does not solve the privacy issues stemming from the information sharing between different ISOEMSs mentioned before.*

For this spatial form of decomposition, all network elements will be assigned to a specific operational zone. Thereby, the following special network elements are defined:

Definition 3.1

- ***Border edge***: *An edge e crossing the border of two neighboring operational zones, e.g. Zone a and Zone b. Note, that thereby this border edge will be defined as a border edge in both neighboring operational zones a and b.*
- ***Border node***: *A node i connected to a border edge e.*
- ***Border loop***: *Border loops are loops l, which have edges e in multiple operational zones.*

Before different approaches are presented which can be used to decompose the NLP problem (3.4) below, a further entity is defined, which is needed to enable a distributed solution. This is the coordinator:

Definition 3.2

The ***coordinator*** *synchronizes calculations and checks if the stopping criterion[a] is fulfilled for every iteration of the distributed optimization approach [HLW17], applied in the design of the ISOEMSs. Note, that in this context synchronizing is understood as starting and ending parallel or sequential calculations of the different entities, in the correct order and point in time.*

[a] Different stopping criteria, are provided in Section 3.3.2, with the respective distributed optimization approaches.

In the following Sections different approaches, which can be used to perform a distributed NLP problem optimization are presented and discussed.

Distributed Optimization Approaches

Several approaches exist which can be used to solve large scale optimization problems through decomposition into smaller subproblems which are then solved in parallel [CCMGB06]. A wide number of papers has been published on the application of such techniques to the OPF problem, in order to find optimal dispatch solutions of large scale EPNs. In doing so, distributed operation is enabled while achieving global optimality [NPC03]. An overview of these approaches applied to different forms of the OPF problem is given in [MDS+17].

OCD and ADMM represent the SOTA and have been successfully applied within the context of optimal dispatch of MESs , as will be stated in detail below. Thus, these two approaches and further developments of these algorithms will be presented in the following four paragraphs on a high level. Thereafter, toward the end of the current section, the aim is to select an appropriate distributed optimization approach for the design of the ISOEMSs within the TCS . This selection is based on a methodical comparison.

Alternating Direction Method of Multipliers The ADMM approach brings together the Dual Decomposition, the Augmented Lagrangian, and the Method of Multipliers methodologies [MDS+17]. The description of the ADMM algorithm and properties below follow [BPC+11, pp. 13-17]. Given the following problem:

$$\min_{\boldsymbol{x},\boldsymbol{z}} f(\boldsymbol{x}) + g(\boldsymbol{z}) \tag{3.27a}$$

subject to

$$\tilde{\boldsymbol{A}}\boldsymbol{x} + \tilde{\boldsymbol{B}}\boldsymbol{z} = \tilde{\mathrm{c}} \tag{3.27b}$$

with optimization variable vectors $\boldsymbol{x}$ and $\boldsymbol{z}$, objective functions f and g, matrices $\tilde{\boldsymbol{A}}$ and $\tilde{\boldsymbol{B}}$, and vector $\tilde{\mathrm{c}}$. The Augmented Lagrangian (AL) $\mathcal{L}^{\mathrm{al}}$ of problem (3.27) is given by:

$$\mathcal{L}^{\mathrm{al}}(\boldsymbol{x},\boldsymbol{z},\boldsymbol{\lambda}) = f(\boldsymbol{x}) + g(\boldsymbol{z}) + \boldsymbol{\lambda}^\top(\tilde{\boldsymbol{A}}\boldsymbol{x} + \tilde{\boldsymbol{B}}\boldsymbol{z} - \tilde{\mathrm{c}}) + (\rho^{\mathrm{al}}/2)||\tilde{\boldsymbol{A}}\boldsymbol{x} + \tilde{\boldsymbol{B}}\boldsymbol{z} - \tilde{\mathrm{c}}||_2^2 \tag{3.28}$$

with the AL penalty parameter ρ^{al} and the dual variable vector $\boldsymbol{\lambda}$. The different ADMM iterations ν are defined as:

$$\boldsymbol{x}_{\nu+1} := \underset{\boldsymbol{x}}{\operatorname{argmin}}\; L_{\rho^{\mathrm{al}}}(\boldsymbol{x},\boldsymbol{z}_\nu,\boldsymbol{\lambda}_\nu) \tag{3.29a}$$

$$\boldsymbol{z}_{\nu+1} := \underset{\boldsymbol{z}}{\operatorname{argmin}}\; L_{\rho^{\mathrm{al}}}(\boldsymbol{x}_{\nu+1},\boldsymbol{z},\boldsymbol{\lambda}_\nu) \tag{3.29b}$$

$$\boldsymbol{\lambda}_{\nu+1} := \boldsymbol{\lambda}_\nu + \rho^{\mathrm{al}}(\tilde{\boldsymbol{A}}\boldsymbol{x}_{\nu+1} + \tilde{\boldsymbol{B}}\boldsymbol{z}_{\nu+1} - \tilde{\mathrm{c}}) \tag{3.29c}$$

Supposing, the two following assumptions hold:

Assumption 3.4. *The functions f and g are proper, convex, and closed.*

Assumption 3.5. *There exists a saddle point on the (unaugmented) Lagrangian $\mathcal{L}$[91].*

Then, the following convergence properties are shown in [BPC+11]: For $\nu \to \infty$ the value of $f(\boldsymbol{x}_\nu) + g(\boldsymbol{z}_\nu)$ will converge to the optimal objective of (3.27), fulfilling (3.27b). Further, $\boldsymbol{\lambda}_\nu$ converges towards the dual optimal vector $\boldsymbol{\lambda}^*$.

91 With $\mathcal{L} = \mathcal{L}^{\mathrm{al}}$ for $\rho^{\mathrm{al}} = 0$.

For practical implementations it is of interest to use a stopping criterion, which defines when the procedure defined in (3.29) can be ended. This is typically checked by the coordinator as described in Definition 3.2. A possible stopping criterion is given in [BPC+11, p. 33].

Within the application of an optimal dispatch problem of energy systems with two ISOEMSs, one for operational Zone a and one for Zone b. Then, the variables $\boldsymbol{x}$ are optimized during the optimization of the ISOEMS in Zone a and similarly, the ISOEMS responsible for Zone b optimizes the variables $\boldsymbol{z}$. Then through (3.27b) a subset of the variables in $\boldsymbol{x}$ and $\boldsymbol{z}$ become equal.

Further properties of the ADMM approach are listed in Table 3.2. Thereby it is important to note, that the basic ADMM approach presented here needs a central coordinator performing the calculation (3.29c) and the calculations in (3.29) are performed sequentially, first by Zone a, then by Zone b and finally by the coordinator. Thus, ADMM in this basic form helps to reduce the problem sizes of the subproblems, but does not enable parallel computation. Further, ADMM in this basic form is no fully distributed approach following Definition 1.1, due to the calculations of the coordinator.

Distributed Approaches of the Alternating Direction Method of Multipliers Different approaches exist, to enable a fully distributed and thus parallel calculation of an ADMM implementation. The main difference to the standard ADMM approach introduced in the last paragraph is that the calculation of $\boldsymbol{\lambda}$ is performed by every operational zone in parallel instead of using a central coordinator. Hence, the role of the coordinator becomes less important and the information exchange between the subproblem solving controllers and the coordinator can be reduced, compare Table 3.2 and Table 3.3. The central basic idea behind these approaches, that makes it possible to use ADMM in a distributed form, is presented here for the case of two operational zones/subproblems. These are Zone a and Zone b notated by the superscripts "za" and "zb" in the following. Supposing subproblems with local variables, here $\boldsymbol{x}^{\mathrm{za}}$ and $\boldsymbol{x}^{\mathrm{zb}}$, the main idea is to introduce global consensus variables $\boldsymbol{z}$ which will be recalculated based on the average values of all local variables in $\boldsymbol{x} = [\boldsymbol{x}^{\mathrm{za}}, \boldsymbol{x}^{\mathrm{zb}}]$, representing the consensus variable locally. Assuming a central problem of the following form, considered as a General Form Consensus Problem (GFCP) in [BPC+11]:

$$\min_{\boldsymbol{x},\boldsymbol{z}} f^{\mathrm{za}}\left(\boldsymbol{x}^{\mathrm{za}}\right) + f^{\mathrm{zb}}\left(\boldsymbol{x}^{\mathrm{zb}}\right) \tag{3.30a}$$

subject to

$$\tilde{\boldsymbol{A}}\boldsymbol{x} = \boldsymbol{z} \tag{3.30b}$$

with $\boldsymbol{z} = [\boldsymbol{z}^{\mathrm{za}}, \boldsymbol{z}^{\mathrm{zb}}]$, matrices $\tilde{\boldsymbol{A}} = \mathrm{diag}(\tilde{\boldsymbol{A}}^{\mathrm{za}}, \tilde{\boldsymbol{A}}^{\mathrm{zb}})$, and the objective functions of both operational zones f^{za} and f^{zb}. Therein, a subset of the local variables in $\boldsymbol{x}$, is represented in both subproblems and thus in $\boldsymbol{x}^{\mathrm{za}}$ and $\boldsymbol{x}^{\mathrm{zb}}$. Therefore, as described before for the general ADMM approach, also in this case, some variables of the initial central optimization problem will be found in both subproblems.

Table 3.2: Properties of the ADMM algorithm.

Property	Description
Convergence Speed	Strongly depends on the choice of ρ^{al}. Several adaptive algorithms are provided in the literature to find appropriate values for different classes of ADMM algorithms [MDS+17].
Form of Decomposition and Coordination in MESs	Over border edges or border nodes of the network graphs, by duplicating the elements, with their respective variables into both adjacent operational zones. Coordination is then achieved through the equality constraint (3.27b) which enforces equality of these border variables, as a subset of the variables $\boldsymbol{x}$ and $\boldsymbol{z}$ of the two subproblems.
Information Exchange during runtime	Zone a sends $\boldsymbol{x}$ to Zone b, Zone b sends $\boldsymbol{z}$ to Zone a, both operational zones send their variables to the coordinator, coordinator sends $\boldsymbol{\lambda}$ to Zone a and b. In this pure form the algorithm is not scheduled in parallel.
Additional role of the coordinator[97]	Calculation of $\boldsymbol{\lambda}$ in every iteration of the ADMM approach.
Empirical Studies	An overview of ADMM applied to EPN dispatch is given in [MDS+17]. Case studies showing the applicability of ADMM to a small scale and large-scale CEPDHN optimization are given in [LWZS17], however in both test systems the ADMM implementation needs at least 2.78 times longer to solve the given problem then the central solver. A distributed DHN operation based on a MILP problem is provided in [KMHG21].

Note, that the general form consensus problem given in (3.30) arises from (3.27), by setting:

$$f(\boldsymbol{x}) = f^{\mathrm{za}}(\boldsymbol{x}^{\mathrm{za}}) + f^{\mathrm{zb}}(\boldsymbol{x}^{\mathrm{zb}}), \qquad g(\boldsymbol{z}) = 0, \qquad \tilde{\boldsymbol{B}} = -\boldsymbol{I}, \qquad \tilde{c} = \boldsymbol{0} \tag{3.31}$$

where $\boldsymbol{I}$ defines the identity matrix.

The distributed ADMM iterations for operational Zone a can then be calculated by:

$$\boldsymbol{x}^{\mathrm{za}}_{\nu+1} := \underset{\boldsymbol{x}^{\mathrm{za}}}{\operatorname{argmin}}\ f^{\mathrm{za}}(\boldsymbol{x}^{\mathrm{za}}) + \boldsymbol{\lambda}^{\mathrm{za}^\top}\left(\tilde{\boldsymbol{A}}^{\mathrm{za}}\boldsymbol{x}^{\mathrm{za}} - \boldsymbol{z}^{\mathrm{za}}_{\nu}\right) + \frac{\rho^{\mathrm{al}}}{2}||\tilde{\boldsymbol{A}}^{\mathrm{za}}\boldsymbol{x}^{\mathrm{za}} - \boldsymbol{z}^{\mathrm{za}}_{\nu}||_2^2 \tag{3.32a}$$

$$\boldsymbol{z}^{\mathrm{za}}_{\nu+1} := \frac{1}{2}(\tilde{\boldsymbol{A}}^{\mathrm{za}}\boldsymbol{x}^{\mathrm{za}}_{\nu+1} + \tilde{\boldsymbol{A}}^{\mathrm{zb}}\boldsymbol{x}^{\mathrm{zb}}_{\nu+1}) \tag{3.32b}$$

$$\boldsymbol{\lambda}^{\mathrm{za}}_{\nu+1} := \boldsymbol{\lambda}^{\mathrm{za}}_{\nu} + \rho^{\mathrm{al}}(\tilde{\boldsymbol{A}}^{\mathrm{za}}\boldsymbol{x}^{\mathrm{za}}_{\nu+1} - \boldsymbol{z}^{\mathrm{za}}_{\nu+1}) \tag{3.32c}$$

thereby, it can be seen, that $\boldsymbol{x}^{\mathrm{za}}_{\nu+1}$ can be directly calculated independently of Zone b. For the calculation of $\boldsymbol{z}^{\mathrm{za}}_{\nu+1}$ in (3.32b) the newly calculated variables of Zone b $\boldsymbol{x}^{\mathrm{zb}}_{\nu+1}$ need to be communicated. The new values of the dual variables $\boldsymbol{\lambda}^{\mathrm{za}}_{\nu+1}$ are then again calculable independently. Note, that the primal and dual variables of Zone b are equally calculated, by replacing all "za" entries in (3.32) by "zb" and vice versa[92].

Further, note, that it is possible to limit $\boldsymbol{x}$ and $\boldsymbol{z}$ in (3.30) and thus also in (3.32a) to $\boldsymbol{x}^{\mathrm{za}} \in \mathcal{X}^{\mathrm{za}}$, $\boldsymbol{x}^{\mathrm{zb}} \in \mathcal{X}^{\mathrm{zb}}$, and $\boldsymbol{z} \in \mathcal{Z}$. Where, the spaces $\mathcal{X}^{\mathrm{za}}$, $\mathcal{X}^{\mathrm{zb}}$, and $\mathcal{Z}$ can be defined by (local) constraints. This enables to e.g. include nonlinear power flow constraints, when this approach is applied to solve distributed optimal dispatch problems, as in [GHT17].

Optimality Condition Decomposition The decomposition approach is presented for the case of two subproblems/operational zones, marked with the superscripts ^{za} and ^{zb} below. An exemplary central optimization problem used to illustrate the approach is given as [CCMGB06]:

$$\min_{\boldsymbol{x}^{\mathrm{za}},\boldsymbol{x}^{\mathrm{zb}}}\ f^{\mathrm{za}}(\boldsymbol{x}^{\mathrm{za}}) + f^{\mathrm{zb}}(\boldsymbol{x}^{\mathrm{zb}}) \tag{3.33a}$$

subject to

$$\boldsymbol{h}^{\mathrm{za}}(\boldsymbol{x}^{\mathrm{za}}, \boldsymbol{x}^{\mathrm{zb}}) = \boldsymbol{0}, \tag{3.33b}$$

$$\boldsymbol{h}^{\mathrm{zb}}(\boldsymbol{x}^{\mathrm{za}}, \boldsymbol{x}^{\mathrm{zb}}) = \boldsymbol{0} \tag{3.33c}$$

with the equality constraints divided into to parts $\boldsymbol{h}^{\mathrm{za}}$ and $\boldsymbol{h}^{\mathrm{za}}$. Then, the Lagrangian is defined as:

$$\begin{aligned}\mathcal{L}(\boldsymbol{x}^{\mathrm{za}}, \boldsymbol{x}^{\mathrm{zb}}, \boldsymbol{\lambda}^{\mathrm{za}}, \boldsymbol{\lambda}^{\mathrm{zb}}) =& f^{\mathrm{za}}(\boldsymbol{x}^{\mathrm{za}}) + f^{\mathrm{zb}}(\boldsymbol{x}^{\mathrm{zb}}) \\ &+ (\boldsymbol{\lambda}^{\mathrm{za}})^\top \boldsymbol{h}^{\mathrm{za}}(\boldsymbol{x}^{\mathrm{za}}, \boldsymbol{x}^{\mathrm{zb}}) + (\boldsymbol{\lambda}^{\mathrm{zb}})^\top \boldsymbol{h}^{\mathrm{zb}}(\boldsymbol{x}^{\mathrm{za}}, \boldsymbol{x}^{\mathrm{zb}})\end{aligned} \tag{3.34}$$

[92] Therefrom it can be seen, that in the case of two operational zones/subproblems the global variables which are of relevance in Zone a and Zone b are the same, as $\boldsymbol{z}^{\mathrm{za}} = \boldsymbol{z}^{\mathrm{zb}}$.

Table 3.3: Properties of the class of distributed ADMM algorithms.

Property	**Description**
Convergence Speed	Strongly depends on the choice of ρ^{al}. Several adaptive algorithms are provided in the literature to find appropriate values for different classes of ADMM algorithms [MDS$^+$17]. The tuning of ρ^{al} represents a tradeoff between faster convergence speed and a remaining gap between the Lagrange multipliers of the central solution and those of the distributed solution [GHT17]. Further, the convergence speed is dependent on the chosen decomposition [GHT15].
Form of Decomposition and Coordination in MESs	Over border edges or border nodes of the network graphs, by duplicating the elements, with their respective variables into both adjacent operational zones, and by additionally introducing global variables $\boldsymbol{x}$ of these duplicated variables in $\boldsymbol{z}$. Coordination is then achieved through the equality constraint (3.30b).
Information Exchange during runtime	Zone a sends $\boldsymbol{x}^{\mathrm{za}}$ to Zone b, and receives $\boldsymbol{x}^{\mathrm{zb}}$ from Zone b. If an adaptive penalty parameter updating method is used, as e.g. presented in [GHT17], then additional communication is necessary for every ADMM iteration ν.
Additional role of the coordinator[97]	-
Empirical Studies	Case studies of this approach applied to the OPF problem show, that these approaches can lead to convenient results in different test systems, including several of the IEEE benchmark networks [MDS$^+$17, GHT17, Ers14, Ers15, MVC17, CKC$^+$14]. Application to MESs are given in [XWZ$^+$19, LGZ$^+$19] for a simplified DHN model (e.g. assuming constant mass flows) and in [YHA$^+$20, XZL$^+$20] without DHN models. Demand response in a DHN is shown in [CYW20], however, the DHN is not decomposed therefore but ADMM is used to coordinate the network dispatch with distributed building agents.

Setting up the Karush-Kuhn-Tucker (KKT) conditions and applying the Newton-Raphson method leads to the following set of equations, that is solved for every Newton iteration ν [Arn11]:

$$\underbrace{\begin{bmatrix} \mathbf{KKT}^{\mathrm{za}} & \mathbf{KKT}^{\mathrm{zb,za}} \\ \mathbf{KKT}^{\mathrm{za,zb}} & \mathbf{KKT}^{\mathrm{zb}} \end{bmatrix}}_{\mathbf{KKT}} \underbrace{\begin{bmatrix} \boldsymbol{\Delta}^{\mathrm{cent,za}} \\ \boldsymbol{\Delta}^{\mathrm{cent,zb}} \end{bmatrix}}_{\boldsymbol{\Delta}^{\mathrm{cent}}} = - \underbrace{\begin{bmatrix} \nabla_{[\boldsymbol{x}^{\mathrm{za}},\boldsymbol{\lambda}^{\mathrm{za}}]}\mathcal{L} \\ \nabla_{[\boldsymbol{x}^{\mathrm{zb}},\boldsymbol{\lambda}^{\mathrm{zb}}]}\mathcal{L} \end{bmatrix}}_{\nabla\mathcal{L}} \tag{3.35}$$

with the differential operator ∇ and the search directions of the central problem given by $\boldsymbol{\Delta}^{\mathrm{cent,za}} = \left[\Delta\boldsymbol{x}^{\mathrm{cent,za}}, \Delta\boldsymbol{\lambda}^{\mathrm{cent,za}}\right]$ and $\boldsymbol{\Delta}^{\mathrm{cent,zb}} = \left[\Delta\boldsymbol{x}^{\mathrm{cent,zb}}, \Delta\boldsymbol{\lambda}^{\mathrm{cent,zb}}\right]$. Further, the KKT matrices are defined as [CCMGB06]:

$$\begin{aligned} \mathbf{KKT}^{\mathrm{za}} &= \begin{bmatrix} \nabla^2_{\boldsymbol{x}^{\mathrm{za}},\boldsymbol{x}^{\mathrm{za}}}\mathcal{L} & \left(\nabla_{\boldsymbol{x}^{\mathrm{za}}}\boldsymbol{h}^{\mathrm{za}}\right)^\top \\ \nabla_{\boldsymbol{x}^{\mathrm{za}}}\boldsymbol{h}^{\mathrm{za}} & \mathbf{0} \end{bmatrix}, & \mathbf{KKT}^{\mathrm{zb,za}} &= \begin{bmatrix} \nabla^2_{\boldsymbol{x}^{\mathrm{zb}},\boldsymbol{x}^{\mathrm{za}}}\mathcal{L} & \left(\nabla_{\boldsymbol{x}^{\mathrm{za}}}\boldsymbol{h}^{\mathrm{zb}}\right)^\top \\ \nabla_{\boldsymbol{x}^{\mathrm{zb}}}\boldsymbol{h}^{\mathrm{za}} & \mathbf{0} \end{bmatrix}, \\ \mathbf{KKT}^{\mathrm{za,zb}} &= \left(\mathbf{KKT}^{\mathrm{zb,za}}\right)^\top, & \mathbf{KKT}^{\mathrm{zb}} &= \begin{bmatrix} \nabla^2_{\boldsymbol{x}^{\mathrm{zb}},\boldsymbol{x}^{\mathrm{zb}}}\mathcal{L} & \left(\nabla_{\boldsymbol{x}^{\mathrm{zb}}}\boldsymbol{h}^{\mathrm{zb}}\right)^\top \\ \nabla_{\boldsymbol{x}^{\mathrm{zb}}}\boldsymbol{h}^{\mathrm{zb}} & \mathbf{0} \end{bmatrix} \end{aligned} \tag{3.36}$$

Applying the OCD approach to (3.33) leads to the following two subproblems, for Zone a:

$$\min_{\boldsymbol{x}^{\mathrm{za}}} f^{\mathrm{za}}\left(\boldsymbol{x}^{\mathrm{za}}\right) + \left(\overline{\boldsymbol{\lambda}}^{\mathrm{zb}}\right)^\top \boldsymbol{h}^{\mathrm{hb}}\left(\boldsymbol{x}^{\mathrm{za}}, \overline{\boldsymbol{x}}^{\mathrm{zb}}\right) \tag{3.37a}$$

subject to

$$\boldsymbol{h}^{\mathrm{za}}\left(\boldsymbol{x}^{\mathrm{za}}, \overline{\boldsymbol{x}}^{\mathrm{zb}}\right) = \mathbf{0} \tag{3.37b}$$

and Zone b:

$$\min_{\boldsymbol{x}^{\mathrm{zb}}} f^{\mathrm{zb}}\left(\boldsymbol{x}^{\mathrm{zb}}\right) + \left(\overline{\boldsymbol{\lambda}}^{\mathrm{za}}\right)^\top \boldsymbol{h}^{\mathrm{za}}\left(\overline{\boldsymbol{x}}^{\mathrm{za}}, \boldsymbol{x}^{\mathrm{zb}}\right) \tag{3.38a}$$

subject to

$$\boldsymbol{h}^{\mathrm{zb}}\left(\overline{\boldsymbol{x}}^{\mathrm{za}}, \boldsymbol{x}^{\mathrm{zb}}\right) = \mathbf{0} \tag{3.38b}$$

where the $\overline{\square}$ operator indicates, that these (dual) variables have been turned into parameters. And all variables and constraints have been assigned to a certain operational zone, while the objective function has been split up into two parts. Further (3.37b) and (3.38b) represent complicating constraints, which are understood as:

Definition 3.3

*A **complicating constraint** is dependent on one or more variables of its proper operational zone and one or more variables from one or more further operational zones. All variable values coming from different operational zones are then turned into parameters by the approach, and are thus treated as constants, updated for every iteration ν.*

It has to be noted, that complicating constraints are not only considered within their proper operational zone by the approach, but also as soft constraints within the objectives of the operational zones, from which the $\overline{\square}$ parameters originate from, when they are used as constraints, see (3.37) and (3.38). In either case, the used $\overline{\square}$ parameters are obtained from the (dual) variables of the other operational zone from the last Newton iteration $v-1$. This indicates the basic OCD algorithm for every Newton iteration ν:

1. Perform one Newton iteration of the two subproblems (3.37) and (3.38).
2. Exchange the information/parameters between the subproblems/operational zones, here $\overline{\boldsymbol{x}}^{\mathrm{za}}$, $\overline{\boldsymbol{x}}^{\mathrm{zb}}$, $\overline{\boldsymbol{\lambda}}^{\mathrm{za}}$, and $\overline{\boldsymbol{\lambda}}^{\mathrm{zb}}$.
3. Check if the stopping criterion is fulfilled. If it is not met yet, calculate the next Newton iteration $v+1$ in step 1.

Different stopping criteria are stated in the given literature. In [CCMGB06, p. 216f] and [Arn11, p. 97] a tolerance ϵ of the (exchanged) optimization variables between two iterations is used, while in [CNP02], also the first order optimality measure is compared against a used tolerance. Computing these distributed subproblems with Newtons method is equal to solving the following set of equations, within every Newton iteration [Arn11]:

$$\underbrace{\begin{bmatrix}\mathbf{KKT}^{\mathrm{za}} & \mathbf{0}\\ \mathbf{0} & \mathbf{KKT}^{\mathrm{zb}}\end{bmatrix}}_{\overline{\mathbf{KKT}}}\underbrace{\begin{bmatrix}\boldsymbol{\Delta}^{\mathrm{dist,za}}\\ \boldsymbol{\Delta}^{\mathrm{dist,zb}}\end{bmatrix}}_{\boldsymbol{\Delta}^{\mathrm{dist}}} = -\begin{bmatrix}\nabla_{[\boldsymbol{x}^{\mathrm{za}},\boldsymbol{\lambda}^{\mathrm{za}}]}\mathcal{L}\\ \nabla_{[\boldsymbol{x}^{\mathrm{zb}},\boldsymbol{\lambda}^{\mathrm{zb}}]}\mathcal{L}\end{bmatrix} \tag{3.39}$$

with the distributed search directions $\boldsymbol{\Delta}^{\mathrm{dist,za}} = \left[\Delta\boldsymbol{x}^{\mathrm{dist,za}}, \Delta\boldsymbol{\lambda}^{\mathrm{dist,za}}\right]$ and $\boldsymbol{\Delta}^{\mathrm{dist,zb}} = \left[\Delta\boldsymbol{x}^{\mathrm{dist,zb}}, \Delta\boldsymbol{\lambda}^{\mathrm{dist,zb}}\right]$ as well as the approximated KKT matrix $\overline{\mathbf{KKT}}$. Note that by comparing (3.39) with (3.35) it is understood that $\overline{\mathbf{KKT}}$ is obtained from $\mathbf{KKT}$ by setting the matrices on the secondary diagonals to $\mathbf{0}$. Now, convergence of the distributed solution towards the central solution of (3.33) can be shown if at the provided second-order KKT point $\boldsymbol{y}^* = \left[\boldsymbol{x}^{\mathrm{za}}, \boldsymbol{x}^{\mathrm{zb}}, \boldsymbol{\lambda}^{\mathrm{za}}, \boldsymbol{\lambda}^{\mathrm{zb}}\right]$ the following holds [CNP02]:

Assumption 3.6. *Lipschitz continuous second-order derivatives of the equations f^{za}, f^{zb}, $\boldsymbol{h}^{\mathrm{za}}$, and $\boldsymbol{h}^{\mathrm{zb}}$ exist.*

Assumption 3.7. *The Jacobian matrix of the equality constraints of* (3.33), *precisely* $\left[(\nabla\boldsymbol{h}^{\mathrm{za}})^\top, (\nabla\boldsymbol{h}^{\mathrm{zb}})^\top\right]^\top$ *provides full row rank.*

Assumption 3.8. *For* (3.33), *the sufficient second-order optimality conditions are fulfilled.*

Further, for every iterate of Newton's method ν, the approximated KKT matrix $\overline{\mathbf{KKT}}_\nu$ has to fulfill the condition below:

Condition 3.1:
All matrices $\overline{\mathbf{KKT}}_\nu$ of the sequence $\overline{\mathbf{KKT}}_\nu \to \overline{\mathbf{KKT}}^*$ are non singular.

Thereby, $\overline{\mathbf{KKT}}^*$ describes the approximated KKT matrix at the optimal solution. Last but not least, the condition for convergence given below needs to hold at $\boldsymbol{y}^*$:

Condition 3.2:

$$\rho^{\mathrm{ocd},*} = \rho\left(\boldsymbol{I} - \left(\overline{\mathbf{KKT}}^*\right)^{-1}\mathbf{KKT}^*\right) < 1, \tag{3.40}$$

with $\rho(\square)$ defining the spectral radius. Further, $\mathbf{KKT}^*$ represents the KKT matrix at the optimal solution. If Assumptions 3.6 to 3.8 are fulfilled, and Condition 3.1 and 3.2 are fulfilled as well, then iteratively solving (3.37) and (3.38) will converge towards $\boldsymbol{y}^*$ with at least linear rate $\rho^{\mathrm{ocd},*}$[93] for all starting points $\boldsymbol{y}^0$ sufficiently close to $\boldsymbol{y}^*$. The literature often refers to $\rho^{\mathrm{ocd},*}$ as the coupling factor. As can be understood from equation (3.40), the coupling factor is smaller for decompositions possessing less coupling, thus less non zero entries on the secondary diagonals of the KKT block matrix $\mathbf{KKT}^{\mathrm{za,zb}}$ and $\mathbf{KKT}^{\mathrm{zb,za}}$. In the context of optimal dispatch of energy systems this refers to less tie lines connecting the decoupled operational zones [Arn11]. In this sense, a high degree of sparsity of the nodal admittance matrix, see Remark 2.2, is a salient property of an EPN, as it increases the probability of finding a very appropriate decomposition.

Further properties of the OCD approach are given in Table 3.4.

Within the following remarks further important aspects of the OCD approach are described.

Remark 3.14:
The OCD approach, which was presented here for the simple case of an optimization problem with complicating constraints, can be simply expanded to the case of additional non complicating equality and inequality constraints, which only include variables of the respective operational zone. Also complicating inequality constraints, as e.g. flow limits over border edges, can be readily included in the same manner as complicating equality constraints [NPC03].

Remark 3.15:
There are indications that symmetrically decomposed problems are more likely to fulfill Condition 3.2. Symmetrical decomposition means, that a complicating constraint in both operational zones is present, which has an identical part and incorporates the same (parameterized) variables. The first indication is, that the majority of the published applications of OCD to energy system dispatch problems decompose the central problems in a symmetrical form. For example in OPF problems, the power flow over a tie line is part of the power flow constraints, see equation (2.15), of both border nodes in the two adjacent operational zones. This power flow is dependent on the voltage amplitude and phase of the bordering nodes, which are thus part of both complicating

[93] Note, that it is important to distinguish between the result of the spectral radius of the matrix defined in Condition 3.2 ρ^{ocd} and the AL tuning parameter ρ^{al} used in the ADMM algorithms. The identical symbols are used here to stay in line with the prevailing amount of literature on both approaches. However, distinguishing superscripts al and ocd are used for clarification.

[94] Strictly regarded, this approach decouples the CEPDHN over EPN tie lines, which are connected to energy converters as EBs and CHPs.

Table 3.4: Properties of the OCD algorithm.

Property	Description
Convergence Speed	At least with linear rate $\rho^{\text{ocd},*}$ [CNP02].
Form of Decomposition and Coordination in MESs	Typically over border edges of the network graphs; thereby, the power flow over the border edges is defined through potential variables of the connected border nodes/buses.
Information Exchange during runtime	The operational zones exchange all variables appearing in the complicating constraints and the respective Lagrange multipliers of the complicating constraints within every OCD iteration.
Additional role of the coordinator[97]	-
Empirical Studies	OCD has been widely tested for optimal dispatch problems for EPNs including several of the IEEE benchmark networks [CNP02, NPC03, GHT15, GHT16, HGA09]. Further work is stated in [MDS+17]. In the context of CEPDHN an approach for the decomposition between EPNs and DHNs is presented in [HLW17][94].

constraints. The second indication is, that in [Ham20] it is shown, that in case of unsymmetrical decomposition of a convex optimization problem, certain parameter constellations of the objective function do not fulfill Condition 3.2[95]*.*

Remark 3.16:

Slight model reformulations can have a large impact on the coupling of the operational zones, and thus, as well on the coupling factor $\rho^{\text{ocd},*}$ but also on the amount of necessary information exchange. A simple example demonstrating a reduction of information exchange is given below in Example 3.2. Also, see Section 3.3.2 for an example on decrease of coupling, in the context of DHN operation. Hence, possible model reformulations should always be checked before decomposing a problem into subproblems for the OCD approach.

[95] Also it appears reasonable, that the distributed search directions $\boldsymbol{\Delta}^{\text{dist}}$ will most likely be closer to the central search direction $\boldsymbol{\Delta}^{\text{cent}}$ if all operational zones take information of the other (bordering) operational zones into account, while calculating $\boldsymbol{\Delta}^{\text{dist}}$.

Example 3.2:

Using the power flow equations of the border buses in the form stated in equations (2.15)*, all bus voltages and phases from Buses* i_2*,* i_3*, and* i_4 *need to be send to the Zone a to apply the OCD algorithm. This would entail, that not only bordering operational zones exchange information but also operational Zones a and c would have to exchange information. In order to prevent this the power flow equation* (2.15) *of e.g. bus* i_2 *can be defined as:*

$$P_{i_2,k} = \sum_{i \in \mathbb{S}_i^{\mathrm{epn}}} P^{\mathrm{flow}}_{i,i_2,k}, \quad \forall\, k \in \mathbb{S}_k \tag{3.41}$$

$$P^{\mathrm{flow}}_{i,i_2,k} = V_{i,k} V_{i_2,k} [B^{\mathrm{bus}}_{i,i_2} \sin(\delta_{i,k} - \delta_{i_2,k}) + G^{\mathrm{bus}}_{i,i_2} \cos(\delta_{i,k} - \delta_{i_2,k})], \quad \forall\, i \in \mathbb{S}_i^{\mathrm{epn}},\ k \in \mathbb{S}_k \tag{3.42}$$

where the flows between buses is defined as P^{flow}*. Using this alternative model will prevent operational Zone c to communicate information to Zone a, during the solution process. Explicitly in this case, only Zone b will communicate the flows* $\overline{P^{\mathrm{flow}}_{i_2,i_2,k}}$*,* $\overline{P^{\mathrm{flow}}_{i_2,i_4,k}}$*, and* $\overline{P^{\mathrm{flow}}_{i_2,i_3,k}}$*, as well as* $V_{i_2,k}$ *and* $\theta_{i_2,k}$ *to Zone a. Note, that these considerations can be similarly applied to the reactive power flow equation* (2.16)*.*

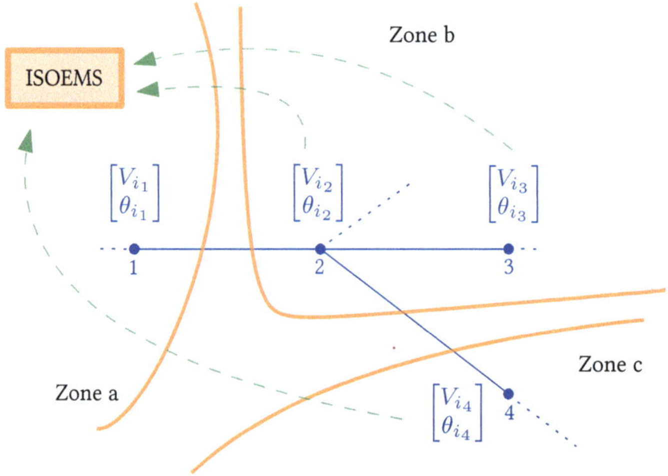

Figure 3.9: Example showing inadequate modeling can lead to more information exchange.

In the following Remark 3.17 a possibility to prevent possible singularity of the approximated KKT matrix $\overline{\mathbf{KKT}}$ is provided.

Remark 3.17:

If not all variables occurring in the optimization problem are included in the objective function $f(\boldsymbol{x})$*, which is the case in the objective* (3.5) *this can result in a rank deficiency of the Hessian of* *the Lagrangian of the central problem* $\nabla^2_{\boldsymbol{x},\boldsymbol{x}}\mathcal{L}$. *The same considerations are valid regarding the distributed subproblems of the OCD approach. Thus, also* $\nabla^2_{\boldsymbol{x}^{\mathrm{za}},\boldsymbol{x}^{\mathrm{za}}}\mathcal{L}$ *and* $\nabla^2_{\boldsymbol{x}^{\mathrm{zb}},\boldsymbol{x}^{\mathrm{zb}}}\mathcal{L}$ *can be rank deficient. This entails that the approximated KKT matrix* $\overline{\mathbf{KKT}}_\nu$ *could become singular for certain Newton iterations* ν*, and thus Condition 3.1 would be violated. In this case, the convergence of the OCD approach towards the central solution is not be guaranteed. To guard against singularity in this case, the Hessian of the Lagrangian is slightly modifiable to* $\nabla^2_{\boldsymbol{x},\boldsymbol{x}}\mathcal{L} + \delta \boldsymbol{I}$*, where* δ *represents a sufficiently small parameter and* $\boldsymbol{I}$ *the identity matrix [NW06, p. 574]. In doing so, the result of the solution is only slightly altered, while singularity of the modified Hessian of the Lagrangian* $\nabla^2_{\boldsymbol{x},\boldsymbol{x}}\mathcal{L} + \delta \boldsymbol{I}$ *can be prevented. This measure is equivalent to extending the central objective* $f(\boldsymbol{x})$ *to:*

$$f^{\mathrm{new}}(\boldsymbol{x}) = f(\boldsymbol{x}) + f^{\delta}(\boldsymbol{x}) \tag{3.43}$$

where the newly added extension $f^{\delta}(\boldsymbol{x})$ *is given by* $f^{\delta}(\boldsymbol{x}) = \delta \boldsymbol{x}^{\top} \boldsymbol{I} \boldsymbol{x}$. *Note, that also the new central objective* $f^{\mathrm{new}}(\boldsymbol{x})$ *is intuitively decomposed into the objectives of the respective operational zones, as* $f^{\delta}(\boldsymbol{x})$ *can be directly decomposed for the subproblems into:*

$$f^{\delta}(\boldsymbol{x}) = f^{\delta,\mathrm{za}}(\boldsymbol{x}) + f^{\delta,\mathrm{zb}}(\boldsymbol{x}) = \delta {\boldsymbol{x}^{\mathrm{za}}}^{\top} \boldsymbol{I} \boldsymbol{x}^{\mathrm{za}} + \delta {\boldsymbol{x}^{\mathrm{zb}}}^{\top} \boldsymbol{I} \boldsymbol{x}^{\mathrm{zb}} \tag{3.44}$$

Remark 3.18:

OCD is known to reduce computation time for increasing problem sizes with adequate decompositions [CNP02, NPC03, GHT16]. Applied to small optimization problems OCD tends to increase the computation time [CNP02, GHT16]. This comes from the fact that OCD usually increases the amount of necessary Newton iterations due to the performed approximation of the KKT matrix by the approach. Nevertheless, the single iterations are usually performed faster with increasing reduction for rising problem sizes. This speed up results for example from the smaller sizes of the KKT matrices of the subproblems, which reduces the necessary computation time for matrix inversions or the solution of linear systems of equations [CNP02].

Optimality Condition Decomposition with Preconditioned Conjugate Gradient

In case Condition 3.2 is not fulfilled, a Preconditioned Conjugate Gradient Method (PCGM) [CNP02] can be applied, to alter the search direction Δ^{dist} in a form that reduces its deviation from Δ^{cent}. This can be achieved by modifying equation (3.39) with an appropriate preconditioner $\boldsymbol{\Pi}$ as stated below [Arn11]:

$$\boldsymbol{\Pi}^{-1}\overline{\mathbf{KKT}}\boldsymbol{\Delta}^{\mathrm{dist,new}} = -\boldsymbol{\Pi}^{-1}\nabla\mathcal{L} \tag{3.45}$$

Then (3.45) is solved approximately for every iteration of the OCD approach using a Generalized Minimal Residual (GMRES) procedure. The central challenge of this approach is to find a good preconditioner $\mathbf{\Pi}$ [CNP02]. Desired features of $\mathbf{\Pi}$ are [Arn11, CNP02]:

1. The preconditioner $\mathbf{\Pi}$ has to provide full rank, as can be understood from (3.45).
2. Its matrix inversion should be numerically inexpensive to prevent high additional computational burden.
3. Further, it should be of such form, that it is sufficiently close to $\overline{\mathbf{KKT}}$[96], while enabling an appropriate transformation so that the newly calculated search direction $\mathbf{\Delta}^{\mathrm{dist,new}}$ will be sufficiently close to $\mathbf{\Delta}^{\mathrm{cent}}$.

Thus, a straightforward choice for the preconditioner is the KKT matrix of the central optimization problem as $\mathbf{\Pi} = \mathbf{KKT}$ [Arn11]. However, the main disadvantage of this selection is the fact, that the necessity to calculate the inversion $\mathbf{KKT}$, which needs to be done by a central coordinator, will inhibit the desired benefits of the OCD approach, as distributed/parallel optimization without the need for a central coordinator. Also the calculation of $\mathbf{KKT}^{-1}$ will become computationally expensive for large scale problems. Thus, other generally applicable preconditioners for nonlinear optimization problems need to be found first, in order to apply this approach effectively [CNP02]. Table 3.5 summarizes the rest of the relevant properties of this approach. Also note that strictly regarded the special OCD algorithm presented here is no fully distributed approach following Definition 1.1, due to the calculations of the coordinator.

Comparison and Selection As stated in [MDS$^+$17] for the EPN only case: "Further empirical work is needed to better characterize the practical performance of distributed optimization algorithms and the selection of appropriate tuning parameters for various power system optimization problems." As the optimization problem derived in this work, regarding the CEPDHN case, with partly new component models, broad empirical studies are not provided by current literature, and thus the decision for one of the given approaches is based on the methodological properties of the four approaches presented in the four previous sections. Therefore, it is assumed, that the assumptions and conditions presented with the respective approaches hold.

The basic ADMM algorithm presented in Section 3.3.2 has several disadvantages over the distributed ADMM , as defined in Section 3.3.2. These are, first, the more sequential character of the calculation, leaving out options of parallelization. Second, the central coordinator needs to calculate the updates of the Lagrange multipliers $\boldsymbol{\lambda}$ in every iteration, leading to additional information exchange.

Similarly OCD with PCGM has the qualitative disadvantages over OCD [98] to need additional calculations by the coordinator, including the inversion of a potentially large matrix, the preconditioner $\mathbf{\Pi}$.

[96] Choosing $\mathbf{\Pi} = \overline{\mathbf{KKT}}$ leads to trivial calculation of $\mathbf{\Delta}^{\mathrm{dist,new}}$ in (3.45).

[97] The term additional is used here, to distinguish between the basic role of the coordinator, as synchronizing calculations and checking the stopping criterion, as defined in Section 3.3.2.

[98] If OCD itself fulfills Condition 3.2.

Table 3.5: Properties of the OCD with PCGM algorithm.

Property	Description
Convergence Speed	At least with linear rate [CNP02].
Form of Decomposition and Coordination in MESs	Typically over border edges of the network graphs; thereby, the power flow over the border edges is defined through potential variables of the connected border nodes/buses.
Information Exchange during runtime	The operational zones exchange all variables appearing in the complicating constraints and the respective lagrange multipliers of the complicating constraints within every OCD iteration. Additionally, all variables necessary for the preconditioner calculation are sent to the coordinator, which then sends the corrected search directions $\boldsymbol{\Delta}^{\mathrm{dist,new}}$ to the individual operational zones.
Additional role of the coordinator[97]	Check if Condition 3.2 is fulfilled. If this is not the case, calculation of the preconditioner matrix $\boldsymbol{\Pi}$.
Empirical Studies	OCD with PCGM has been tested for optimal dispatch problems for EPNs including several of the IEEE benchmark networks [CNP02, NPC03, GHT15, GHT16, HGA09]. Further work is stated in [MDS$^+$17]. In [Arn11], this approach is applied to the decomposition of EPNs and gas networks in MESs .

Thus the most promising approaches for the application here are OCD and the distributed ADMM approaches. Both algorithms enable mostly parallel calculations and a very low amount of information exchange. Still, OCD has two qualitative advantages over ADMM :

1. The OCD approach does not duplicate variables into the adjacent operational zones, and into the global variables, through the form of decomposition. This becomes especially relevant when not only a single time step k is optimized but instead several time steps $|\mathbb{S}_k|$ of the moving horizon approach are taken into account.
2. Further ADMM necessitates the tuning of the AL penalty parameter ρ^{al}. Indeed the literature states several approaches for adaptive calculation of ρ^{al} within every ADMM iteration ν, but these approaches again build upon the tuning of new parameters [Ers15, GHT17]. The design of these novel coefficients, does not only influence the convergence speed, but also possibly distinguishes between convergence (to the local optimum) and divergence [MVC17]. Also it has to be kept in mind, that the tuning of ρ^{al} represents a tradeoff between convergence speed and a remaining dual gap of the marginal values of the central to those of the distributed solution [GHT17][99].

Further, the necessity of the OCD approach to fulfill Condition 3.2 does not decisively need to be understood as a limitation of the approach. As a central question in the design of distributed optimal dispatch approaches is always on how the operational zones should be chosen, which is closely related to how the decomposition should be performed. This is of high relevance as it has direct impact on the convergence speed of the distributed optimization approach [GHT15]. Therefore, Condition 3.2 directly serves as an additional quantitative criterion for optimal cell dimensioning, as shown in [GHT15, GHT16].

Based on the above reasons, the OCD approach will be used in this work to implement the distributed ISOEMSs.

However, so far no literature exists, that has shown the effective application of the OCD approach to optimal dispatch problems in DHNs. Thus, a new approach is developed here with the aim of bridging that gap, to enable a decomposition of the overall optimal CEPDHNs dispatch problem.

[99] Thus, possibly impeding LMPAs.

Distributed ISOEMS Design

In this Section the design of the distributed ISOEMS is presented. In general it is very similar to the design of the central ISOEMS. The main difference is, that the calculation of the optimal control values, based on solving the operational NLP problem (3.4), is performed in a distributed form. Note, that this is the second step during run time shown in the Nassi-Shneidermann diagram in Figure 3.7. In the distributed case, a coordinator, as introduced in Definition 3.2, will synchronize the distributed optimization, performed in parallel by the different ISOEMSs. Also the coordinator will end the distributed calculation as soon as a predefined stopping criterion is fulfilled.

As discussed in Section 3.3.2, the OCD approach is selected to perform the distributed optimization within this work. The OCD algorithm has already been used to decompose EPNs into different zones, and to divide CEPDHNs into EPNs and DHNs, which are then optimized in a parallel form, see Table 3.4. A decomposition of DHNs, has been contributed within this work [MIJSH22], and is presented in the following.

Distributed Optimal Operation of District Heating Networks based on Optimality Condition Decomposition The following shows how optimal operation problems for DHNs, can be solved in parallel, applying the OCD approach to the central optimization problem of an ISOEMS (3.4). Parts of this section have been published in [MIJSH22, pp. 1-18]. Note, that (3.4) describes the NLP solved by a central ISOEMS to optimally operate an entire CEPDHN. In this Section however, the description is restricted to the case where an ISOEMS runs a DHN and thus all parts of the objective function $\boldsymbol{f}$, the equality constraints $\boldsymbol{h}$, or the inequality constraints $\boldsymbol{g}$ are left aside. When referring to (3.4) within this Section, only the parts of the problem which are part of a DHN optimal operation problem are meant. As will be shown in this section, applying OCD to DHNs comes along with several difficulties, which do not occur when the distributed optimization approach is used to parallelize the solution procedure of EPNs. A first difficulty is to find a form of decomposition that leaves enough degrees of freedom within the system of equations, of every subproblem. This is demonstrated in the introducing Example 3.3. Due to the strong interdependence of flows and pressure drops of the symmetric supply and return network, and due to the presence of hydraulic node and loop equations, it is non trivial to consider the main hydraulic equations, while finding a form of decomposition that leaves enough degrees of freedom. This is a central difference to applying OCD to the OPF for optimal dispatch of EPNs, where the subproblems always have sufficient degrees of freedom, see Example 3.4.

Example 3.3:

We consider the stationary hydraulic model, of the network shown in Figure 3.10, to demonstrate the challenge arising from applying OCD to DHNs. A general optimization problem (3.46) *is assumed here, where the objective of the central problem* (3.46a) *is represented as the sum of the objectives the two subproblems f^{za} and f^{zb}. The equations* (3.46b) *and* (3.46b) *represent the hydraulic node and loop models which have been introduced in Chapter 2 in* (2.23) *and* (2.24)[100]*. Besides* f^{hydr} *in* (3.46d) *represents the nonlinear relation between mass flow* $\dot{m}$ *and differential pressure* Δp *on an edge* e*.*

$$\min f^{\text{za}} + f^{\text{zb}} \tag{3.46a}$$

w.r.t.

$$\sum_{e \in \mathbb{S}_e^{\text{dhn}}} A_{i,e}\dot{m}_e = 0, \quad \forall\, i \in \mathbb{S}_i^{\text{dhn}} \tag{3.46b}$$

$$\sum_{e \in \mathbb{S}_e^{\text{dhn}}} B_{l,e}\Delta p_e = 0, \quad \forall\, l \in \mathbb{S}_l^{\text{dhn}} \tag{3.46c}$$

$$\Delta p_e - f^{\text{hydr}}(\dot{m}_e) = 0, \quad \forall\, e \in \mathbb{S}_e^{\text{dhn}} \tag{3.46d}$$

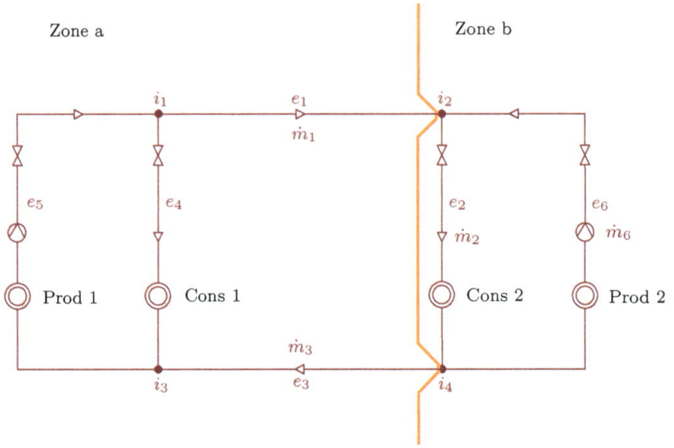

Figure 3.10: Example DHN decomposed into Zone a and Zone b.

A straightforward form of decomposition based on OCD can be performed by decoupling between node i_2 *and edge* e_1 *and between node* i_4 *and edge* e_3*, as delineated in Figure 3.10. By assigning the border loop*[101] *equation of* l_2 *to Zone b, the complicating constraints, see Definition 3.3, of Zone b are given by:*

$$\overline{\dot{m}_1} - \dot{m}_2 + \dot{m}_6 = 0 \tag{3.47}$$

$$\dot{m}_2 - \overline{\dot{m}_3} - \dot{m}_6 = 0 \tag{3.48}$$

$$\overline{\Delta p_1} + \Delta p_2 + \overline{\Delta p_3} - \overline{\Delta p_4} = 0 \tag{3.49}$$

The value of Δp_2 *is determined through equation* (3.49)*. Based on* (3.46d) *the mass flow* $\dot{m}_2$ *is also determined in this case, e.g. assuming that Edge* e_2 *represents a valve with a fixed valve flow factor* K*, and the constant component coefficient* μ_2 *for Edge* e_2 *is obtained from aggregation as described in Remark 3.19 below. Also the other variables of edge* e_6*, precisely* Δp_6 *and* $\dot{m}_6$*, are known in this scenario through the respective equations in* (3.46)*. Thus, in* *this instance, the solution of the optimization problem of Zone b has zero degrees of freedom, and will in this case by defined by the fixed variable values* $\overline{\square}$ *obtained from Zone a. In this case, the distributed convergence properties will most likely be negatively affected as the calculations of Zone b only have a limited effect, providing the new variables* $\boldsymbol{x}^{\mathrm{zb}}$ *and Lagrange multipliers* $\boldsymbol{\lambda}^{\mathrm{zb}}$ *used in the soft constraints of Zone a, see* (3.37a). *Intuitively speaking, Zone b cannot actively affect the overall search direction of both zones but provide information to Zone a, how the search direction calculated by Zone a will affect the objective of Zone b in every Newton iteration* ν. *If both zones have zero degree of freedom, see Appendix B for an example, the distributed solution of the OCD approach will definitively not equal the solution of the central problem* (3.46), *except for the trivial case, where the problem is initialized at the optimal solution.*

In summary, this example shows, that it is not always reasonable to apply OCD, due to a lack of degrees of freedom within the system of equations, of the equality constraints of the subproblems. However, it is still possible to do so in the aforementioned case, but either the convergence speed may reduced if one subproblem has zero degrees of freedom or the solution of the OCD approach will be identical to the initial values of the optimization variables $\boldsymbol{x}_0$, *if both zones have zero degrees of freedom.*

Example 3.4:

In contrast to DHNs, applying the OCD approach to the optimal dispatch of EPNs cannot reduce the degrees of freedom of the subproblems to zero. This can be seen with regard to the simple meshed example EPN shown in Figure 3.11. The resulting complicating constraint for Bus i_1 *in Zone a results from* (2.15) *and is given by:*

$$\begin{aligned} P_{i_1} = \; & V_{i_1} V_{i_2} [B^{\mathrm{bus}}_{i_1,i_2} \sin(\delta_{i_1} - \delta_{i_2}) + G^{\mathrm{bus}}_{i_1,i_2} \cos(\delta_{i_1} - \delta_{i_2})] \\ & + V_{i_1} \overline{V}_{i_3} [B^{\mathrm{bus}}_{i_1,i_3} \sin(\delta_{i_1} - \overline{\delta}_{i_3}) + G^{\mathrm{bus}}_{i_1,i_3} \cos(\delta_{i_1} - \overline{\delta}_{i_3})] \\ & + V_{i_1}^2 G_{i_1,i_1} \end{aligned} \tag{3.50}$$

As can be seen from (3.50) *also within the complicating constraints the bus voltage amplitude and angle, here* V_{i_1} *and* δ_{i_1} *of Bus* i_1 *remain variable, including the power flow over* e_1[102]. *As the same applies to* V_{i_2} *and* δ_{i_2} *in the complicating constraint for Bus* i_2 *this shows, that the subproblem of Zone a also has sufficient degrees of freedom when applying the OCD approach.*

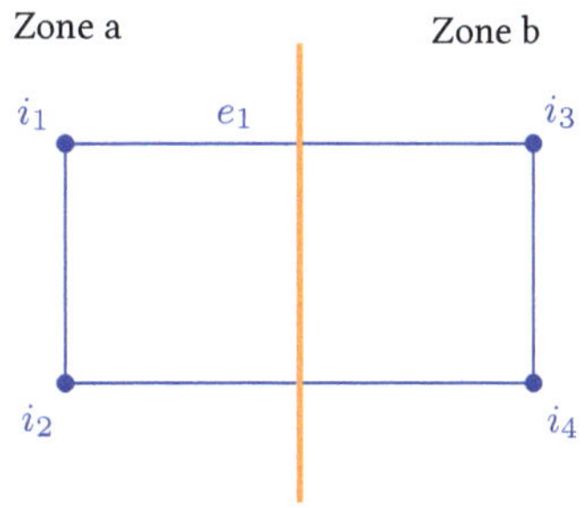

Figure 3.11: Example EPN decomposed into Zone a and Zone b.

The presented Examples 3.3 and 3.4 demonstrated, that applying OCD to DHNs can simply lead to a system of equations of the subproblems with insufficient degrees of freedom, while this problem does not arise, when OCD is used for distributed optimization of EPNs.

Remark 3.19:

The model equations and the therein used variables of multiple network components can be aggregated into a single network component to reduce the amount of optimization variables and the implementation burden to test new approaches for small DHNs as in Example 3.3, see also [Ill20]. For example the pressure drop over Edge e_5 *between Nodes* i_1 *and* i_3 *can be written as:*

$$\Delta p_5 = p_3 - p_1 = \beta_5 - \mu_5(\dot{m}_5)^2 \tag{3.51}$$

where the variable component coefficients β_5 *and* μ_5 *each represent a sum of the respective component coefficients. Thus for example* β_5 *can be written as the sum of the component coefficient of the valve* β_5^{vlv}*, the pump* β_5^{pump} *and the heat exchanger* β_5^{exch} *resulting in* $\beta_5 = \beta_5^{\text{vlv}} + \beta_5^{\text{pump}} + \beta_5^{\text{exch}}$*. This is based on the formulas for* β *and* μ *of different model components in Section 2.3.2. An analogous formula can be set up for* μ_5*. Note, that taking into account, that the same differential pressure* Δp_5 *can occur for different valve and pump set points, this degree of freedom can be used to further aggregate* (3.51) *into* $\Delta p_5 = p_3 - p_1 = \beta_5$ *eliminating the degree of freedom. Note, that also the thermal component models of heat exchangers, valves, and pumps in edges* e_2, e_4, e_5, *and* e_6 *can simply be aggregated, as for pumps and valves the output temperature is equivalent to the input temperature, see Section 2.3.3.*

Besides, the need to find a form of decomposition that leaves enough degrees of freedom within the subproblems, also Condition 3.2 is easily violated when OCD is applied to DHNs. This comes from the fact, that decomposing the hydraulic optimization problem (3.46) of a DHN as shown in Figure 3.10 into two similarly sized[103] subproblems directly results in three constraints that represent complicating constraints. Two of these arise from the hydraulic border node equations, stemming from (2.23), which are (3.47) and (3.48) in the provided Example 3.3. The third complicating constraint originates from the border loop equation, coming from (2.24) which is (3.49) in the given Example 3.3. Note, that three complicating constraints of this form also arise if one of the other three possible forms of decomposition is used, see Appendix B.

These special constraints comprehend variables coming from a different subproblem, which are thus treated as parameters during the Netwon iterations of the respective subproblem, see Definition 3.3. If the central optimization problem is extended to a thermal model, as within this application, an additional complicating constraint stems from the thermal node equation (2.47)[104]. In general it is known, that a stronger coupling of the subproblems, which is increased by every complicating constraint, leads to higher values of $\rho^{\text{ocd},*}$ [Arn11], see the paragraph on OCD in Section 3.3.2. Thus, the more complicating constraints exist, the more likely it is,

[103] Note, that a similar size of the subproblems is a desired property as this increases the probability, that both subproblems need a similar calculation time for every Newton iteration ν.

[104] Based on the model presented in Chapter 2 the pipeline model equation (2.66) would also have to be taken into account as a complicating constraint. However, this can either be prevented by using the elimination of variables technique [NW06, p. 426] or through Assumption 3.10, which enables to replace the thermal pipeline model by an identity function mapping the input temperature to the output temperature.

that Condition 3.2 is not met. In contrast to DHNs, applying OCD to EPNs only leads to a single complicating constraint for every border node.

Also Example 3.3 does not enable a symmetric form of decomposition, which can be advantageous as described above, see Remark 3.15. This example shows, that the characteristics of DHNs, including symmetric return and supply networks, need caution while applying the OCD approach in order to enable a symmetric solution, enough degrees of freedom and a low form of coupling between the subproblems.

The approach presented here, to overcome these challenges is based on three key aspects, which are outlined here and presented in detail below:

1. An equivalent reformulation of the optimization problem is used, neglecting redundant variables and equations. This helps to reduce the size of the optimization problem and the coupling between the zones.
2. Based on the aforementioned reformulation a different form of decomposition is performed. Instead of decomposing between network components, as performed in Example 3.3, the decomposition is applied over edges/pipelines, and thus between two nodes.
3. The central optimization problem is extended, by replacing one equality constraint obtained from the model reformulation by two new ones. This enables to obtain symmetrical subproblems by applying the decomposition described above. This is desired, as symmetrical decompositions potentially lead to better convergence properties as depicted in Remark 3.15.

In summary these three measures enable to decompose the new central optimization problem in a symmetrical form while fulfilling Condition 3.2, as it will be demonstrated in Section 4.5.

An exact description of the three aspects is given in the following, starting with the equivalent reformulation of the optimization problem (3.4) solved by a central ISOEMS as described in Section 3.3.1. Therefore, in the following all necessary changes performed, leading to modeling deviations from the central problem (3.4) are stated. To ease the understanding, these model changes and there outplay are described by using the simple DHN shown in Figure 4.22. This DHN is very similar to the one shown in Example 3.3, also taking into account the aggregated model description discussed in Remark 3.19. In order to reduce the coupling of the resulting subproblems two measures are performed: First, the hydraulic differential pressures Δp_e over edges e are replace by pressure potentials p_i at network nodes i by using (2.22). Note, that this relation can be given for the entire DHN as $-{\boldsymbol{A}^{\mathrm{dhn}}}^\top \boldsymbol{p} = \Delta \boldsymbol{p}$ [MCS20]. Inserting this into (2.24), written as $\boldsymbol{B}\Delta\boldsymbol{p} = \boldsymbol{0}$ for the total network, leads to $-\boldsymbol{B}({\boldsymbol{A}^{\mathrm{dhn}}}^\top \boldsymbol{p}) = \boldsymbol{0}$ [Des69]. As $\boldsymbol{B}{\boldsymbol{A}^{\mathrm{dhn}}}^\top = \boldsymbol{0}$ in all cases, the hydraulic loop equation (2.24) can be neglected in the following. Additionally, the hydraulic node equations (2.23) of the border nodes in the return network, i_3 and i_4 in Figure 4.22, are omitted. This measure strongly reduces the coupling of the subproblems, as two complicating constraints of the form (3.54) introduced and explained later on, are eliminated. However, by neglecting the hydraulic loop equations and the hydraulic border nodes of the return networks the model becomes incomplete, as the relation between mass flows and differential pressure of supply and return network on the border edges is no more incorporated in the model. This gap is bridged by explicitly introducing the equality $\Delta p_1 = \Delta p_3$ for Example 3.3 which can be notated as $p_1 - p_2 = p_4 - p_3$ or sorted by zones as $p_1 + p_3 = p_4 + p_2$ into the model. Note, that this also implies $\dot{m}_1 = \dot{m}_3$, based on (2.34). In general this can be notated as:

$$\begin{aligned}&\left(\sum_{e \in \mathbb{S}_e^{\text{dhn,za}}, e \notin \mathbb{S}_e^{\text{dhn,be,za}}} \sum_{i \in \mathbb{S}_i^{\text{dhn,bi,za}}} |B_{l,e}||A_{i,e}^{\text{dhn}}|p_{i,k}\right)\\ &= \left(\sum_{e \in \mathbb{S}_e^{\text{dhn,zb}}, e \notin \mathbb{S}_e^{\text{dhn,be,zb}}} \sum_{i \in \mathbb{S}_i^{\text{dhn,bi,zb}}} |B_{l,e}||A_{i,e}^{\text{dhn}}|p_{i,k}\right),\\ &\forall\, l \in \mathbb{S}_l^{\text{dhn,bl}}, k \in \mathbb{S}_k \end{aligned} \tag{3.52}$$

where the border loops are given in $\mathbb{S}_l^{\text{dhn,bl}}$ and the border edges of Zone a $\mathbb{S}_e^{\text{dhn,be,za}}$ represent a subset of the edges of Zone a given in $\mathbb{S}_e^{\text{dhn,za}}$, and the same notation is used for edges of Zone b equivalently. If more then four nodes are found in the regarded border loop l, the relevant border nodes are found through $|B_{l,e}||A_{i,e}^{\text{dhn}}| = 1$ by checking this condition for all border nodes $i \in \mathbb{S}_i^{\text{dhn,bi}}$ and edges e which are no border edges[105] $e \in \mathbb{S}_e^{\text{dhn}}, e \notin \mathbb{S}_e^{\text{dhn,be}}$ of the respective zone[106].

Based on the aforementioned, the second key aspect of the approach, a different form of decomposing the central problem, is introduced. This means, the decomposition is applied over edges/pipelines and thus between two nodes, as shown in Figure 4.22, instead of decomposing between network components as performed in Example 3.3.

The third key aspect represents the extension of the optimization problem (3.4), by replacing the equality constraint (3.52), obtained from the model reformulation, by two new ones. This extension is used to enable a symmetrical form of decomposition in the following. Applying OCD to (3.46) incorporating the suggested changes stated above, the complicating constraint arising from equation (3.52) is assigned either to Zone a or Zone b, thus leading to an unsymmetrical decomposition. Also, representing the exact same equation (3.52) twice in the central problem would lead to an undesired rank deficiency of the KKT matrix. This can be overcome by replacing (3.52) through the following two equations:

$$\begin{aligned}&\left(\sum_{e \in \mathbb{S}_e^{\text{dhn,za}}, e \notin \mathbb{S}_e^{\text{dhn,be,za}}} \sum_{i \in \mathbb{S}_i^{\text{dhn,bi,za}}} |B_{l,e}||A_{i,e}^{\text{dhn}}|p_{i,k}\right) = p^{\text{pre}},\\ &\left(\sum_{e \in \mathbb{S}_e^{\text{dhn,zb}}, e \notin \mathbb{S}_e^{\text{dhn,be,zb}}} \sum_{i \in \mathbb{S}_i^{\text{dhn,bi,zb}}} |B_{l,e}||A_{i,e}^{\text{dhn}}|p_{i,k}\right) = p^{\text{pre}},\\ &\forall\, l \in \mathbb{S}_l^{\text{dhn,bl}}, k \in \mathbb{S}_k \end{aligned} \tag{3.53}$$

A comparison of these equations (3.53) with the initial form (3.52) shows that thereby the pressure potentials p_i of the nodes in the bordering zone have been replaced by the constant

105 All border edges are given in $e \in \mathbb{S}_e^{\text{dhn,be}}$.

106 Note, that e.g. the left side of (3.52) can be simplified to $\sum_{i \in \mathbb{S}_i^{\text{dhn,bi,za}}} p_{i,k}$ if only one point of coupling/border edge (in the return and supply network) to another zone, exists as in Example 3.3.

predefined pressure value p^{pre}. As p^{pre} is equal in both equations of (3.53), every solution fulfilling (3.53), will also satisfy equation (3.52), as (3.53) is a special case of (3.52). The benefits of this new formulation are that the two equations in (3.53) are naturally assigned to their respective zones, without creating further complicating constraints and thus reducing the coupling of the subproblems. Additionally, this procedure yields symmetrically decomposed subproblems. A further form which can be used to explain the derivation of (3.53) is given in the Appendix in Section C. Besides, the directions of the border edges are slightly varied as explained in the following remark.

Remark 3.20:

The decomposition only becomes fully symmetrical in the resulting border node equations (3.54), (3.59), (3.62), *and* (3.64) *outlayed below, by changing the edge directions of the border edges. As shown in Figure 4.22, edge e_1 is directed away from node i_1 and i_2, leading to $A^{\mathrm{dhn}}_{i_1,e_1} = -1$ and $A^{\mathrm{dhn}}_{i_2,e_1} = -1$. Analogously, e_3 is directed towards i_3 and i_4. Visually speaking, every zone calculates its solution with its own border edge directions, where the supply network border edge is directed away from the border node of the respective zone and the return network edge is directed to the proper border node of this zone. This is a deviation from the standard notation of the directed graph spanning up the DHN, where every edge is directed away from the start node and towards the end node. Still, the resulting system of equations is equivalent, as the differential pressure over the border edges Δp_e is defined using absolute values of $A^{\mathrm{dhn}}_{i,e}$, see* (3.56).

The following paragraphs now first introduce and discuss used assumptions, before the complicating constraints and the objective of the OCD approach are explicitly stated one after the other.

The presented approach is based on the following assumptions, used for simplicity reasons:

Assumption 3.9. *Pipelines with VMFDs are the only component types for which edges can become border edges. These pipelines are not part of a mesh in the return or supply network. Also, no control paths reach across zone borders.*

These assumptions limit the necessity to list possibly resulting complicating constraints for different component types. For example pumps, that can be operated in both flow directions, would also represent appropriate component models for a decomposition based on the pump models, presented in Section 2.3[107]. Besides, pipelines are the most common components in DHNs, thus the limitations in finding possible decompositions for this assumption are manageable. And as discussed after the upcoming assumption, the edges over which the approach decomposes the DHN can also be virtual pipelines.

Assumption 3.10. *The pipelines, representing the border edges, are regarded as negligibly short, and thus transmission delay, temperature losses, and pressure potential differences due to height differences Δh are neglected here, as $\Delta h = 0$ is assumed.*

[107] Note, that this would imply that a pump is installed in the return and the supply network accordingly.

If this assumption would lead to unwanted deviations, due to a non negligible length of the pipeline, two artificial nodes and pipelines, one in the return and one in the supply network, could be introduced into the model in one of the zones at the end of the current border edge. Then, all parameters of the old border edge pipelines are kept the same, and the newly inserted pipelines, which are set as the new border edges, have the same parameters except the pipeline length L which is chosen as negligently short.

For simplicity and without loss of generality the approach is presented for the case of two zones here. Further, for notational brevity only the optimization problem of Zone a is fully stated, while the NLP of Zone b is directly evolvable thereof. All equations except the ones replaced above, the complicating constraints and the new objectives stated here, stay the same as given in (3.4) for every area.

The new hydraulic node equation, replacing (2.23) for supply network border nodes in Zone a, is given as[108]:

$$h_{i,k}^{\mathrm{dhn,hydr,sn,bi,za}} = \sum_{e \in \mathbb{S}_{\mathrm{e}}^{\mathrm{dhn}}, e \notin \mathbb{S}_{\mathrm{e}}^{\mathrm{dhn,be}}} A_{i,e}^{\mathrm{dhn}} \dot{m}_{e,k} + \sum_{e \in \mathbb{S}_{\mathrm{e}}^{\mathrm{dhn,be}}} A_{i,e}^{\mathrm{dhn}} \dot{m}_{e,k} = 0, \quad \forall\, i \in \mathbb{S}_i^{\mathrm{dhn,sn,bi,za}}, k \in \mathbb{S}_k \tag{3.54}$$

The main difference here to (2.23) is that border edges $e \in \mathbb{S}_{\mathrm{e}}^{\mathrm{dhn,be}}$ and "normal" edges $e \in \mathbb{S}_{\mathrm{e}}^{\mathrm{dhn}}, e \notin \mathbb{S}_{\mathrm{e}}^{\mathrm{dhn,be}}$ are distinguished. Border edge mass flows can then be expressed in dependence of pressure potentials p as stated below. Thereby, taking into account Assumption 3.10, which entails $\Delta h = 0$ and thus $\beta_{e,k} = 0$ in equation (2.34):

$$\begin{aligned} \dot{m}_{e,k} &\approx \mathrm{sgn}_{\Delta\varepsilon}(\Delta p_{e,k}) \sqrt{\frac{|\Delta p_{e,k}|_{\Delta\varepsilon}}{\mu_{e,k}(\dot{m}_{e,k})}} \\ &= \frac{1}{\sqrt{\mu_{e,k}(\dot{m}_{e,k})}} \frac{\Delta p_{e,k}}{\sqrt{|\Delta p_{e,k}|_{\Delta\varepsilon}}}, \qquad \forall\, e \in \mathbb{S}_e^{\mathrm{dhn,be,za}}, k \in \mathbb{S}_k \end{aligned} \tag{3.55}$$

The dependency of the pressure coefficient on the border edge mass flow $\mu_{e,k}(\dot{m}_{e,k})$ is eliminated by utilizing a precalculated pressure coefficient $\mu_{e,k}^{\mathrm{pre}}$, based on the border edge mass flow of the previous optimization, thus using a similar procedure as effectively applied for the thermal pipeline model, recalculating model parameters when applying the rolling horizon, see Step 6 of Figure 3.7. Further, the differential pressure $\Delta p_{e,k}$ over the considered border edge e can be replaced by the difference of the pressure potential of the two connected border nodes i, from which one is in Zone a and the other in Zone b. Thus, the differential pressure over the border edge connected to a border node i of Zone a can be written as:

$$\Delta p_{e,k} = \sum_{i \in \mathbb{S}_i^{\mathrm{dhn,bi,za}}} \left(|A_{i,e}^{\mathrm{dhn}}| p_{i,k} - \sum_{i' \neq i, i' \in \mathbb{S}_i^{\mathrm{dhn}}} |A_{i',e}^{\mathrm{dhn}}| \overline{p_{i',k}} \right), \quad \forall\, e \in \mathbb{S}_e^{\mathrm{dhn,be,za}}, k \in \mathbb{S}_k \tag{3.56}$$

108 The extensive superscripts of h stand for: district heating network, hydraulic, supply network, border node, and Zone a.

By inserting (3.56) in (3.55) and exchanging the variable $\mu_{e,k}(\dot{m}_{e,k})$ by the parameter $\mu_{e,k}^{\text{pre}}$, the mass flow on the border edge e itself is given as:

$$\dot{m}_{e,k} \approx \frac{1}{\sqrt{\mu_{e,k}^{\text{pre}}}} \frac{\sum_{i\in\mathbb{S}_i^{\text{dhn,bi,za}}} \left(|A_{i,e}^{\text{dhn}}|p_{i,k} - \sum_{i'\neq i,i'\in\mathbb{S}_i^{\text{dhn}}} |A_{i',e}^{\text{dhn}}|\overline{p_{i',k}}\right)}{\sqrt{\left|\sum_{i\in\mathbb{S}_i^{\text{dhn,bi,za}}} \left(|A_{i,e}^{\text{dhn}}|p_{i,k} - \sum_{i'\neq i,i'\in\mathbb{S}_i^{\text{dhn}}} |A_{i',e}^{\text{dhn}}|\overline{p_{i',k}}\right)\right|_{\Delta\varepsilon}}},$$
$$\forall\, e \in \mathbb{S}_e^{\text{dhn,be,za}}, k \in \mathbb{S}_k \qquad (3.57)$$

Exemplary, for the network shown in Figure 4.22, equation (3.56) is given by $\Delta p_{e_1,k} = p_{i_1,k} - p_{i_2,k}$ for Edge e_1 in Zone a. Based on this, the border edge mass flow $\dot{m}_{e_1,k}$ for Zone a stemming from equation (3.57) results in:

$$\dot{m}_{e_1,k} \approx \frac{1}{\sqrt{\mu_{e_1,k}^{\text{pre}}}} \frac{p_{i_1,k} - p_{i_2,k}}{\sqrt{\left|p_{i_1,k} - p_{i_2,k}\right|_{\Delta\varepsilon}}}, \quad \forall\, k \in \mathbb{S}_k \qquad (3.58)$$

The thermal node equation, replacing (2.47) for border nodes, is formulated as follows:

$$\begin{aligned}
&h_{i,k}^{\text{dhn,therm,bi,za}} = \\
&\Bigg[\sum_{e\in\mathbb{S}_e^{\text{dhn,za}}, e\notin\mathbb{S}_e^{\text{dhn,za,be}}} A_{i,e}^{\text{dhn},-} \max\left(\dot{m}_{e,k}, 0\right) \\
&\quad + \sum_{e\in\mathbb{S}_e^{\text{dhn,za}}, e\notin\mathbb{S}_e^{\text{dhn,za,be}}} A_{i,e}^{\text{dhn},+} \max\left(-\dot{m}_{e,k}, 0\right) \\
&\quad + \sum_{e\in\mathbb{S}_e^{\text{dhn,za,be}}} A_{i,e}^{\text{dhn},-} \max\left(\dot{m}_{e,k}, 0\right) \\
&\quad + \sum_{e\in\mathbb{S}_e^{\text{dhn,za,be}}} A_{i,e}^{\text{dhn},+} \max\left(-\dot{m}_{e,k}, 0\right)\Bigg] T_{i,k} \\
&\quad - \Bigg[\sum_{e\in\mathbb{S}_e^{\text{dhn,za}}, e\notin\mathbb{S}_e^{\text{dhn,za,be}}} A_{i,e}^{\text{dhn},+} \max\left(\dot{m}_{e,k}, 0\right) T_{e,k}^{\text{out}} \\
&\quad + \sum_{e\in\mathbb{S}_e^{\text{dhn,za}}, e\notin\mathbb{S}_e^{\text{dhn,za,be}}} A_{i,e}^{\text{dhn},-} \max\left(-\dot{m}_{e,k}, 0\right) T_{e,k}^{\text{in}} \\
&\quad + \sum_{e\in\mathbb{S}_e^{\text{dhn,za,be}}} \Bigg(\left(A_{i,e}^{\text{dhn},+} \max\left(\dot{m}_{e,k}, 0\right) + A_{i,e}^{\text{dhn},-} \max\left(-\dot{m}_{e,k}, 0\right)\right) \\
&\quad \sum_{i'\neq i, i'\in\mathbb{S}_i^{\text{dhn}}} |A_{i,e}^{\text{dhn}}||A_{i',e}^{\text{dhn}}|\overline{T_{i',k}}\Bigg)\Bigg] = 0, \\
&\forall\, i \in \mathbb{S}_i^{\text{dhn,bi,za}}, k \in \mathbb{S}_k
\end{aligned} \qquad (3.59)$$

Note, that similarly as in (3.54) the border mass flows are also defined by (3.57) in (3.59) above. With Assumption 3.10, supposing negligibly short pipelines between zones, the thermal pipeline model reduces to an identity function mapping the input temperature to the output temperature, here $T_{i,k} = \overline{T}_{i',k}$[109]. For Node i_1 in Zone a in the DHN in Figure 4.22 equation (3.59) is written as:

$$\begin{aligned}
&\Big[\max\big(\dot{m}_{e_1,k},0\big)+\max\big(\dot{m}_{e_4,k},0\big)+\max\big(-\dot{m}_{e_5,k},0\big)\Big]T_{i_1,k}\\
&-\Big[\max\big(-\dot{m}_{e_4,k},0\big)T^{\text{in}}_{e_4,k}+\max\big(\dot{m}_{e_5,k},0\big)T^{\text{out}}_{e_5,k}+\max\big(-\dot{m}_{e_1,k},0\big)\overline{T_{i_2,k}}\Big]=0,\\
&\forall\, k\in\mathbb{S}_k
\end{aligned} \tag{3.60}$$

The resulting objective function for Zone a $f^{\text{dhn,za}}$, is defined as given below:

$$\begin{aligned}
f^{\text{dhn,za}} =& f^{\text{dhn,rest,za}} + \sum_{k\in\mathbb{S}_k}\Bigg[\sum_{i\in\mathbb{S}_i^{\text{dhn,sn,bi,zb}}}\Big(\lambda_{i,k}^{\text{dhn,hydr,sn,bi,zb}} h_{i,k}^{\text{obj,dhn,hydr,sn,bi,zb}}\Big)\\
&+\sum_{i\in\mathbb{S}_i^{\text{dhn,bi,zb}}}\Big(\lambda_{i,k}^{\text{dhn,therm,bi,zb}} h_{i,k}^{\text{obj,dhn,therm,bi,zb}}\Big)\Bigg]
\end{aligned} \tag{3.61}$$

This objective function $f^{\text{dhn,za}}$ is composed of two parts. The first describes the decomposed objective of a central ISOEMS $f(\boldsymbol{x})$ explained in Remark 3.17, which is defined here as $f^{\text{dhn,rest,za}}$ for Zone a. Note, that applying the approach described in Remark 3.17 helps to guard against singularity of the approximated KKT matrix $\overline{\mathbf{KKT}}$, while simultaneously pursuing the general objective of an ISOEMS, as stated in Section 3.3.1. The second part is constituted by the sum of the products of the Lagrange multipliers λ and the complicating constraints h coming from Zone b. As indicated by the superscripts the Lagrange multipliers come from Zone b of the last OCD iteration from the complicating constraints $h_{i,k}^{\text{dhn,hydr,sn,bi,zb}}$ and $h_{i,k}^{\text{dhn,therm,bi,zb}}$ of the respective border nodes. The complicating constraints itself are utilized in the slight modified form given below in (3.62) - (3.64) with the additional $\texttt{obj}$ superscript. In comparison to (3.55), (3.57), and (3.59) all variables become OCD parameters and vice versa. Also all sets which have been based on Zone a now become dependent on Zone b and the other way around. This leads to the following definitions:

$$\begin{aligned}
h_{i,k}^{\text{obj,dhn,hydr,sn,bi,zb}} =& \sum_{e\in\mathbb{S}_e^{\text{dhn}},e\notin\mathbb{S}_e^{\text{dhn,be}}} A_{i,e}^{\text{dhn}}\overline{\dot{m}}_{e,k} + \sum_{e\in\mathbb{S}_e^{\text{dhn,be}}} A_{i,e}^{\text{dhn}}\dot{m}_{e,k} = 0,\\
&\forall\, i\in\mathbb{S}_i^{\text{dhn,sn,bi,z}},\, k\in\mathbb{S}_k
\end{aligned} \tag{3.62}$$

[109] However, in theory more detailed models as e.g. a static pipeline model similar to (2.71) or a symmetric dynamic thermal pipeline model could also be considered in the thermal node equation (3.59).

where the border edge mass flow is defined similarly as in (3.57) by:

$$\dot{m}_{e,k} \approx \frac{1}{\sqrt{\mu^{\text{pre}}_{e,k}}} \frac{\sum_{i\in\mathbb{S}_i^{\text{dhn,bi,zb}}} \left(|A^{\text{dhn}}_{i,e}|\overline{p}_{i,k} - \sum_{i'\neq i\in\mathbb{S}_i^{\text{dhn}}} |A^{\text{dhn}}_{i',e}|p_{i',k}\right)}{\sqrt{\left|\sum_{i\in\mathbb{S}_i^{\text{dhn,bi,zb}}} \left(|A^{\text{dhn}}_{i,e}|\overline{p}_{i,k} - \sum_{i'\neq i\in\mathbb{S}_i^{\text{dhn}}} |A^{\text{dhn}}_{i',e}|p_{i',k}\right)\right|_{\Delta\varepsilon}}},$$

$$\forall\, e \in \mathbb{S}_e^{\text{dhn,be,zb}}, k \in \mathbb{S}_k \qquad (3.63)$$

The thermal node equation is formulated as follows for border nodes:

$$\begin{aligned}
&h^{\text{obj,dhn,therm,bi,zb}}_{i,k} = \\
&\Bigg[\sum_{e\in\mathbb{S}_e^{\text{dhn,zb}}, e\notin\mathbb{S}_e^{\text{dhn,zb,be}}} A^{\text{dhn},-}_{i,e} \max\left(\overline{\dot{m}}_{e,k}, 0\right) \\
&+ \sum_{e\in\mathbb{S}_e^{\text{dhn,zb}}, e\notin\mathbb{S}_e^{\text{dhn,zb,be}}} A^{\text{dhn},+}_{i,e} \max\left(-\overline{\dot{m}}_{e,k}, 0\right) \\
&+ \sum_{e\in\mathbb{S}_e^{\text{dhn,zb,be}}} A^{\text{dhn},-}_{i,e} \max\left(\dot{m}_{e,k}, 0\right) \\
&+ \sum_{e\in\mathbb{S}_e^{\text{dhn,zb,be}}} A^{\text{dhn},+}_{i,e} \max\left(-\dot{m}_{e,k}, 0\right)\Bigg]\overline{T}_{i,k} \\
&- \Bigg[\sum_{e\in\mathbb{S}_e^{\text{dhn,zb}}, e\notin\mathbb{S}_e^{\text{dhn,zb,be}}} A^{\text{dhn},+}_{i,e} \max\left(\overline{\dot{m}}_{e,k}, 0\right)\overline{T^{\text{out}}_{e,k}} \\
&\quad + \sum_{e\in\mathbb{S}_e^{\text{dhn,zb}}, e\notin\mathbb{S}_e^{\text{dhn,zb,be}}} A^{\text{dhn},-}_{i,e} \max\left(-\overline{\dot{m}}_{e,k}, 0\right)\overline{T^{\text{in}}_{e,k}} \\
&\quad + \sum_{e\in\mathbb{S}_e^{\text{dhn,zb,be}}} \Bigg(\left(A^{\text{dhn},+}_{i,e} \max\left(\dot{m}_{e,k}, 0\right) + A^{\text{dhn},-}_{i,e} \max\left(-\dot{m}_{e,k}, 0\right)\right) \\
&\quad \sum_{i'\neq i, i'\in\mathbb{S}_i^{\text{dhn}}} |A^{\text{dhn}}_{i,e}||A^{\text{dhn}}_{i',e}|T_{i',k}\Bigg)\Bigg] = 0, \\
&\forall\, i \in \mathbb{S}_i^{\text{dhn,bi,zb}}, k \in \mathbb{S}_k
\end{aligned} \qquad (3.64)$$

Note, that therein the border mass flows are defined by (3.63).

In summary, the approach is fully outlined. Simulational results will show the limited range of applicability of the presented methodology in Section 4.5.

3.4 Summary and Discussion

The design of a new TCS for CEPDHNs was presented within this chapter. This system is based on the coordinated interaction of multiple EMSSAs, each bidding for a single FNP, and the ISOEMSs, coordinating the FNPs within their operational zone. For this system, relevant market and control mechanisms which are at the core of the TCS have been outlaid and discussed in this chapter. The used control mechanism is a distributed rolling horizon approach, where the distributed optimization problems are to be solved in every time step based on the OCD algorithm. The model-based control approach uses the contributed CEPDHN model presented in Chapter 2. The social welfare is used as the allocation objective for electricity and heat markets, which are cleared in a coordinated form by the TCS, thereby maximizing the welfare of the overall system. Market clearing prices are determined by the HPA, which combines the uniform marginal price obtained from dispatch[110] with paid-as-bid and pay-as-bid for a fraction of the bids. The exact procedure taking place during transactive control of the regarded CEPDHNs is provided in Figure 3.7 for the central case with one ISOEMS. In the case of multiple distributed ISOEMSs, the second step is solved by the OCD approach in a distributed parallelized form. The TCS is designed to achieve welfare maximization, technical efficiency and secure[8] CEPDHN operation and forms Contribution 1 of this work.

In this context, it was shown how important it is to design market and control mechanisms in a joint manner, due to their strong interdependence. These interrelations of the mechanisms can on the one side limit the applicability of certain approaches, and on the other side increase the performance of the TCS if adequately employed. Examples for limiting aspects in the mechanism designs are the disadvantages arising from the use of the LMPA and ADMM in a joint form, or the fact that the LMPA brings along two different LMPs for every FNP in the DHN models used within the rolling horizon approach. These show, that designing market and control mechanisms independently can have serious consequences, as for example imprecise[111] LMPs or slow convergence [GHT17]. This would entail the inability of the TCS to calculate new control values in real time. For credit postings, the interconnected design of market and control mechanisms greatly play out within the ISOEMSs design. For that matter, the bids are provided for the entire prediction horizon, and merged with further data as e.g. predictions on power supply and demand. Additionally, network models and operational constraints of the CEPDHNs are applied within a (distributed) rolling horizon control approach to maximize the welfare of all FNPs. In this sense, allocation and control value determination becomes a single procedure. Also, the simultaneous market clearing of the heat and electric energy markets by the distributed controllers, the ISOEMSs, enables to enhance the overall welfare. Besides, the coordinated design of market and control mechanisms even goes down to far more technical aspects. For example, the definition of optimal operation points of pumps and valves has a direct impact on the social welfare. In this regard, it is very important to mention, that the current literature on comparable approaches usually uses strongly simplified network models, as stated in the Research Gap 1 a) and b) in Section 1.2.5. However, when using a model-based control approach, it is clear that modeling decisions will directly impact the control values and thereby the market outcome. For example predefining the flow direction in the edges of

[110] Defined, as the intersection of the dispatched consumer and producer bid and offer curves.

[111] The lack of accuracy of the LMPs originates from the need of additional iterations, which are very time consuming, while the primal variables are already at or near to the exact values in this case, see Table 3.3 and [GHT17].

a DHN model will have a large impact on the obtainable welfare. Similarly, neglecting the possibilities of pipeline storage or TESS application will directly impact the obtained welfare. Besides the aforementioned, salient technical features of the proposed approach are that the rolling horizon approach will always try to minimize temperatures in the DHNs as fast, and as far as possible in order to minimize losses and thus maximize the welfare. The collection of these examples shows the importance of the joint design of market and control mechanisms for CEPDHNs.

However, achieving welfare maximization, technical efficiency and secure[8] CEPDHN operation does not only rely on the joint design of market and control mechanisms. Two further aspects need to be taken into account here. The first is the fact that the coordinated operation of EPNs and DHNs or MES in general leads to a better overall system performance regarding technical and economic indexes [ZWW+21, WYJ+16, LSF14, DLS+19]. The second aspect is the ability of the presented TCS to merge all possible information on future power supply and demand as well as the respective bid and offer prices, in order to predictively optimally dispatch all FNPs connected to the CEPDHNs.

To the best of the authors knowledge for the distributed rolling horizon approach used in this work, which could be classified as a *nonlinear distributed economic time-variant hybrid discrete model predictive control approach* with varying amount of system states, see Section 3.3 no methodological concept for the proof of Lyapunov or bounded input bounded output stability of the closed-loop case is provided in the literature so far. However, these forms of stability are not perused by transactive control approaches in real-world applications. Much more, desired system behavior is understood as fulfilling operational constraints in order to enable trade, meeting contract commitments of power supply, without causing system damage or unnecessary fatigue.

Note that, the presented TCS enables privacy [KW16] preserving information sharing to the FNPs, as only the bids and offers are communicated to the ISOEMS and no further information[112] needs to be shared, see [LLJZ20], [KW16]. Further, considering information sharing, the following should be noted. In Germany, today the DSO by itself, operates the electric power, gas, water, and district heating systems. For jurisdictions, where there are multiple entities responsible for utility services of electricity and heat, there has to be information sharing agreements between the entities in order to implement the proposed TCS. This will require some coordination and policy formation at the regulatory/municipal governmental levels.

The developed HPA used in the TCS prevents uplift costs due to redispatch. Basing the pricing procedure on uniform marginal pricing achieves incentive compatibly if no FNP can take advantage of its market power and if no network constraint limits the solution set. The later necessitates to price some bids and offers by pay-as-bid or paid-as-bid analogously, as described in Section 3.2.3. This leads to a reduction of the overall system welfare. Still, the optimizer inherently minimizes the additional costs arising within the HPA, in order to maximize the social welfare. This is a central advantage of the HPA in comparison with a standard uniform marginal pricing approach, as the only theoretical limitation of possible uplift costs within the UMPA arises from the amount and size of the bids and offers provided. Besides these aspects, avoiding the necessary command and control actions, to organize the redispatch, arising from

[112] As for example room temperatures, SOCs of battery storages in electrical vehicles or available devices.

a UMPA [MSC03] enhances the security of the network operation, as unforeseen complicated operation conditions are less likely.

The TCS can enable scalability in large scale CEPDHNs through the utilization of the OCD approach, which is reported to reduce calculation times of central optimization problems effectively for EPNs [NPC03]. Hence, enabling parallel computation of smaller subproblems of the large scale nonlinear central operational optimization problem (3.4), as it can be decomposed into subproblems between tie lines of the EPN. Thereby, the conditions stated with the OCD approach always need to be fulfilled. This especially applies for the coupling condition. As simulation results provided in Section 4.5 will show, so far the contributed OCD approach for DHNs does not enable to decompose the overall CEPDHN network graph within DHNs. Earlier relevant work in this field using OCD for electricity and gas networks is provided in [Arn11]. However, therein the design of appropriate market mechanisms were not regarded.

The overall operational optimization problem (3.4) which is solved by the central ISOEMS or the distributed ISOEMSs in the TCS is a nonlinear non-convex optimization problem. Thus, convergence properties can only be evaluated through simulations, as performed in the following Chapter 4. Still, nonlinear non-convex models and thus optimization problems are used in relevant real world applications, due to the absence of reasonable alternatives, as for example in the OPF problem [MDS+17] which is strongly related to (3.4).

Note, that within this chapter it is depicted how the combination of Contributions 2, and 3 jointly bring up the main Contribution 1 of this work, the TCS for CEPDHNs, see Figure 1.2. This is achieved, by outlaying the integration of the detailed CEPDHN model, Contribution 2, into the TCS. Further, the HPA for simultaneous market clearing of interconnected DHN and EPN markets is described, representing Contribution 3 of this work. Additionally, an approach for the distributed optimal dispatch of DHNs, constituting Contribution 4 was developed.

4 Results

Simulation results from five CEPDHNs are presented and discussed within this section to show and evaluate different functionalities, properties, advantages and limitations of the TCS designed in Chapter 3. The main topics elaborated in the following case studies and sections are ordered in the same form as the approaches are introduced in Chapter 3. Therefore, Case Study I will discuss the hybrid pricing approach, Case Study II presents the implementation of a centralized ISOEMS, Case Study III highlights DHN operation with VMFDs and thermal energy storage systems, Case Study IV investigates calculation times in large real world CEPDHNs, and finally Case Study V demonstrates the application of the distributed DHN optimization based on OCD. Parts of this chapter have been published in [MGR$^+$21], [MTK$^+$22], and [MIJSH22].

4.1 Case Study I: Hybrid Pricing Approach

The application of the HPA is demonstrated within this case study. After a description of the regarded CEPDHN and the bids and offers of the FNPs, the results are presented. Note, that parts of this Case Study have been initially presented in [MGR$^+$21, pp. 9-11].

4.1.1 Description

To validate the approach, a version of the case study of the Barry Island from [Liu13, LWJB16] is implemented, as displayed in Figure 4.1. Parameters not stated here are given in [Liu13]. The used network model consists of 32 DHN nodes and eight EPN buses at $11.5\,\text{kV}$. For simplicity and without loss of generality, the underlying radial $0.433\,\text{kV}$ power networks are neglected and their loads are aggregated into the flexible loads at the Buses I, III, IV, V, and VI. Also, a stationary heat propagation model within the pipelines is used. Further, a single time step is optimized, see [MGR$^+$21] for the exact used model. None of the simplifications listed above change the functionality of the approach; much more this helps to ease understanding. Bus II of the EPN is modeled as the reference bus with fixed voltage angle and magnitude. The reactive power consumption of all consumers is modeled by a typical value of $\cos(\varphi) = 0.95$. The impedance values of the transmission lines are scaled up in order to show the integration of the transmission losses. The bids and offer of all FNPs are listed in Table 4.1. Further, this Table also displays the nodes and buses to which an FNP is connected. Hereby, it has to be noted, that the cardinality of the network participants is kept low, for simplicity reasons. Still, it is assumed that no market participant is able to take advantage of his market power. Prices are defined as general monetary units, as they vary strongly in time and different regions depending on developments as regulations, renewable energy sources, fuel prices and CO_2-certificate prices [Kel13].

The COP of the CHP and the HP are modeled with 1.3 and 3.5, see (2.17)[113], as in [Liu13]. It is assumed that no MP takes advantage of his potential market power.

The resulting network topology is displayed in Figure 4.1.

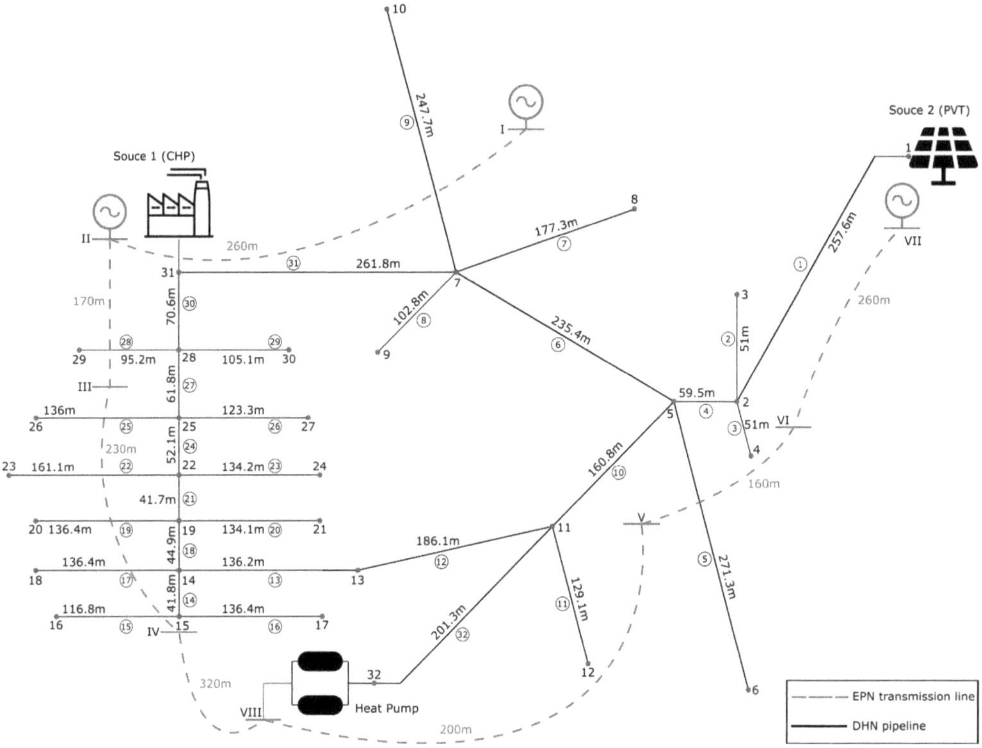

Figure 4.1: Modified Barry Island case study [LWJB16][114].

Typical voltage magnitude limits of $V^{\min} = 0.9$ p.u. and $V^{\max} = 1.1$ p.u. are used. The apparent power flows are limited by $S^{\min} = 0$ p.u. and $S^{\max} = \sqrt{5}S^{\text{base}}$ with $S^{\text{base}} = 1$ kVA, as given in [MGR+21]. The reactive power limits $Q_n^{\min} = -50$ p.u. and $Q_n^{\max} = 50$ p.u. are set for the producers. The node temperature limits are defined by $T_i^{\min} = 40\,°\text{C}$ and $T_i^{\max} = 60\,°\text{C}$. Additionally, the consumer and producer mass flow limits are set to $\dot{m}^{\min} = 0\,\text{kg/s}$ and $\dot{m}^{\max} = 5\,\text{kg/s}$. The output temperatures of all heat consumers are assumed to be $T^{\text{out}} = 30\,°\text{C}$, following typical connecting conditions for 4th generation DHNs to assure low losses [LSG+17]. For simplicity reasons, the return network is assumed to be lossless due to the low temperature gap of 5 °C between the ambient temperature $T^{\text{a}} = 25\,°\text{C}$ and the output

[113] For the CHP /HP the exact values are $a_{n^{\text{epn}},n^{\text{dhn}},1} = 1.3/a_{n^{\text{epn}},n^{\text{dhn}},1} = 3.5$ and $a_{n^{\text{epn}},n^{\text{dhn}},m} = 0$ for all $m \neq 1$.

[114] Credit for this Figure goes to Armin Golla.

Table 4.1: Relevant parameters of all used network participants. The bid and offer costs are stated in general cost units ¤ per unit produced or consumed.

$n \in \mathbb{S}_n^{\text{epn}}$	Type	$i \in \mathbb{S}_i^{\text{epn}}$	P^{min} in p. u.	P^{max} in p. u.	c in ¤/p. u.
1	Consumer	I	-15	0	4.00
2	Consumer	II	-15	0	3.70
3	Consumer	III	-15	0	3.40
4	Consumer	IV	-15	0	3.10
5	Consumer	V	-15	0	2.80
6	Consumer	VI	-15	0	2.50
7	Consumer	VII	-15	0	2.20
8	Heat pump	VIII	-15	0	1.90
9	Power plant	I	0	50	2.50
10	CHP plant	II	0	50	1.50
11	Photovoltaic plant	VII	0	50	1.20
$n \in \mathbb{S}_n^{\text{dhn}}$	Type	$i \in \mathbb{S}_i^{\text{dhn}}$	Φ^{min} in p. u.	Φ^{max} in p. u.	c in ¤/p. u.
1	Consumer	3	-15	0	1.50
2	Consumer	4	-15	0	1.75
3	Consumer	6	-15	0	2.00
4	Consumer	8	-15	0	2.25
5	Consumer	9	-15	0	2.50
6	Consumer	10	-15	0	2.75
7	Consumer	12	-15	0	3.00
8	Consumer	16	-15	0	3.25
9	Consumer	17	-15	0	3.50
10	Consumer	18	-15	0	3.75
11	Consumer	19	-15	0	4.00
12	Consumer	20	-15	0	4.25
13	Consumer	21	-15	0	4.50
14	Consumer	23	-15	0	4.75
15	Consumer	24	-15	0	5.00
16	Consumer	26	-15	0	5.25
17	Consumer	27	-15	0	5.50
18	Consumer	29	-15	0	5.75
19	Consumer	30	-15	0	6.00
20	Solarthermal plant	1	0	90	2.00
21	CHP plant	31	0	65.0	3.00
22	Heat pump	32	0	52.5	1.00

temperature of the consumers. The parameter for the approximation of the maximum and absolute value function (2.20) is set to $\Delta\varepsilon = 1\text{e} - 5$. The marginal costs of the losses in both networks are determined by the UMP , as explained in Remark 3.5. The respective values are $c^{\text{loss,epn}} = 2.2$ ¤/p.u. and $c^{\text{loss,dhn}} = 3$ ¤/p.u., see Figure 4.2 and Figure 4.3.

The case study is implemented in GAMS using the IPOPT solver [NW06]. Thereby, an interior-point method is used to solve the resulting nonlinear programming problem. The solution is calculated in 1.49 s with an Intel i7-6600U CPU @ 2.60 GHz.

4.1.2 Results and Analysis

The aim of this case study is to present the functionality of the approach. The main results are found in Figure 4.2 and Figure 4.3. The submitted and dispatched offers and bids of the producers and consumers are sorted by merit order and separated for the EPN and the DHN. As mentioned in earlier in Remark 3.6, this enables to determine the UMPDs of both networks separately, as they represent two different price zones, which is helpful when the price levels in the different networks deviate strongly. In the depicted scenario, the UMPD in the EPN is $c^{\text{umpd,epn}} = 1.9$ ¤/p.u. while the UMPD in the DHN reaches $c^{\text{umpd,dhn}} = 3$ ¤/p.u..

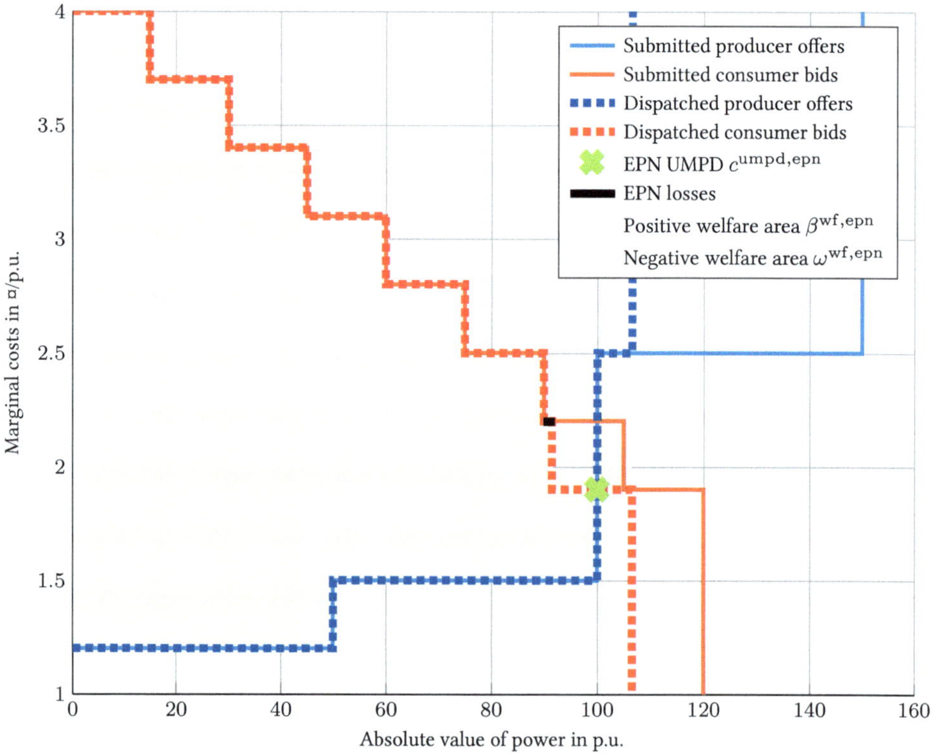

Figure 4.2: Submitted and dispatched offers and bids of electric producers and consumers sorted by merit order and the resulting UMPD for the EPN.

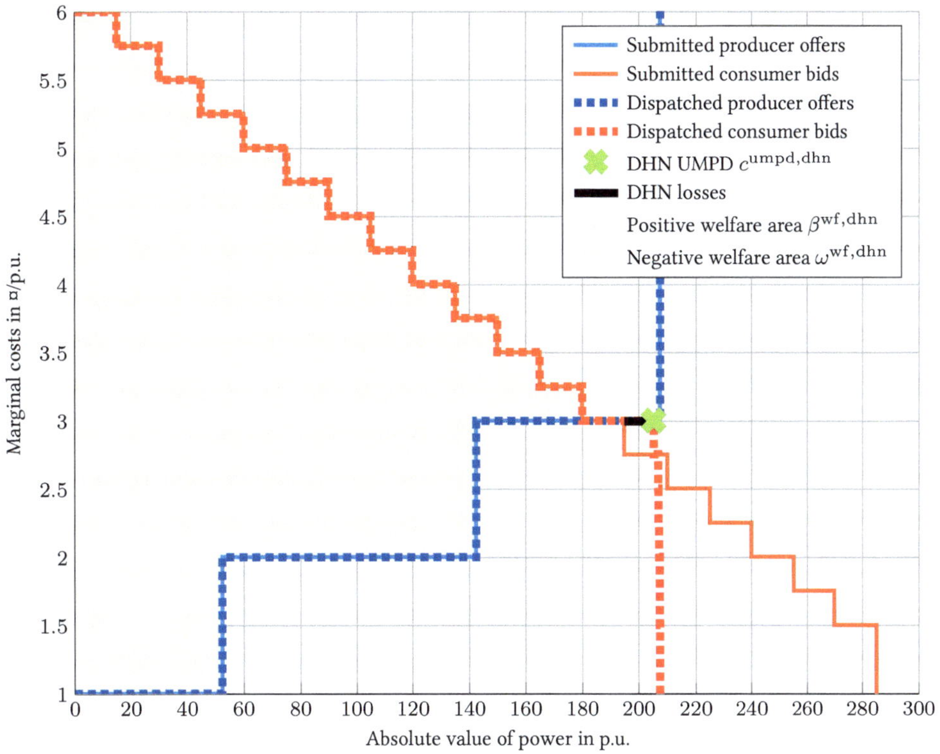

Figure 4.3: Submitted and dispatched offers and bids of heat producers and consumers sorted by merit order and the resulting UMPD for the DHN.

Figure 4.2 and 4.3 show that the approach functions as expected. The solution of the optimization problem shows that the bids of the consumers who are willing to pay the most and the offers of the producers which are willing to produce for the lowest prices are dispatched both in the heat and the electricity network. The power of the electric power plant, which has marginal costs of 2.5 ¤/p.u., as described in Table 4.1, is additionally dispatched. This does not happen if the EPN is optimized on its own. By doing so, parts of the total electric power produced by the power plant, the CHP unit and the photovoltaic plant can be used to fully operate the heat pump (costs of 1.9 ¤/p.u.). Based on this, additional welfare is created in the DHN due to the high COP of the HP of 3.5, which enables the HP operator to offer low price heat power to the DHN consumers. The cost of the respective offer is 1.0 ¤/p.u., see Table 4.1. The resulting uniform price for the EPN $c^{\text{umpd,epn}} = 1.9$ ¤/p.u. is defined by the intersection of the marginal cost curves of the dispatched supply and demand. The real power losses of the EPN add up to $P^{\text{loss,epn}} = 1.5$ p.u. and are also depicted in Figure 4.2. Note that these are plotted as part of the dispatched solution but obviously not part of the plotted submitted bids, as the losses are determined during runtime of the optimization and not known in advance. As the costs of network losses are shared by all grid participants through a grid fee, they can be seen as a part of the welfare. In this understanding, the costs of the network losses have been added to the bids of the consumers in advance. Hereby it is assumed, that producers would directly pass through these grid fee costs to the consumers by raising their offer prices respectively. Then, the entire welfare created in the EPN sums up to $W^{\mathrm{epn}} = \beta^{\mathrm{wf,epn}} - \omega^{\mathrm{wf,epn}} = 173.03$ ¤.

The heat losses which add up to $\Phi^{\mathrm{loss,dhn}} = 10.29$ p.u. are priced with $c^{\mathrm{loss,dhn}} = 3.0$ ¤/p.u. which is equal to the resulting UMPD of the DHN $c^{\mathrm{umpd,dhn}}$, see Figure 4.3. All heat producers are fully dispatched. The consumers with the least priced bids, more exactly heat consumers 1-6, see Table 4.1, are barely dispatched. Based on this, the welfare of the DHN of $W^{\mathrm{dhn}} = 486.56$ ¤ is created. The total welfare created throughout both networks is equal to $W^{\mathrm{tot}} = W^{\mathrm{epn}} + W^{\mathrm{dhn}} = 659.59$ ¤.

All network and operational constraints are fulfilled by the solution of the optimization problem. Therefore, redispatch becomes unnecessary, as requested. Still, some operational constraints are at the provided bounds, as e.g. the node temperatures of the DHN T_i, which vary between $T_1 = 59.99$ °C and $T_9 = 40.00$ °C close and on the bounds defined earlier.

The case study shows that social welfare maximization is provided while secure[8] network operation can be granted for a real network. Also it demonstrates, that the solution of the optimization problem provides small negative welfare areas ω while providing high accumulated total welfare W^{tot}, close to the potential maximum of the unconstrained network case.

4.2 Case Study II: Transactive Control System with Central ISOEMS

The performance of the proposed TC approach is studied considering two scenarios, which represent diverse functionalities of the presented TCS. These results have been previously published in [MTK+22]. The first scenario highlights how flexibility is included in CEPDHN operation by utilizing a BESS, demand side management of consumers in both networks, and the heat storage capabilities of the DHN. The second scenario examines how the operation of a CEPDHN changes for a very cold winter day with high infeed of RESs, thereby demonstrating how the coupled operation improves overall system operation in contrast to independent operation of the networks by a TCS, as reported in [LLJZ20]. The application of price signals based on the HPA or the LMPA for the EPN and the DHN are also discussed.

4.2.1 Scenario 1: DHN Flexibility Provision for the EPN

Description

Electric Power Network The 11 kV distribution network from the case study in [Liu13] is used here as part of the test system, and is shown in Figure 4.4. The EPN comprises ten buses and nine radial feeder sections. The flexible consumers, including the heat pumps, are located at Buses 2, 3, 4, 5, 7, 8, and 9. The producers and energy converters are chosen to clearly illustrate the energy flows in a future energy network. The CHP plant represents a flexible producer whereas, the WPP is modeled as an inflexible producer, which aligns with the German regulation for maximum infeed of RESs. The distribution grid is assumed to have

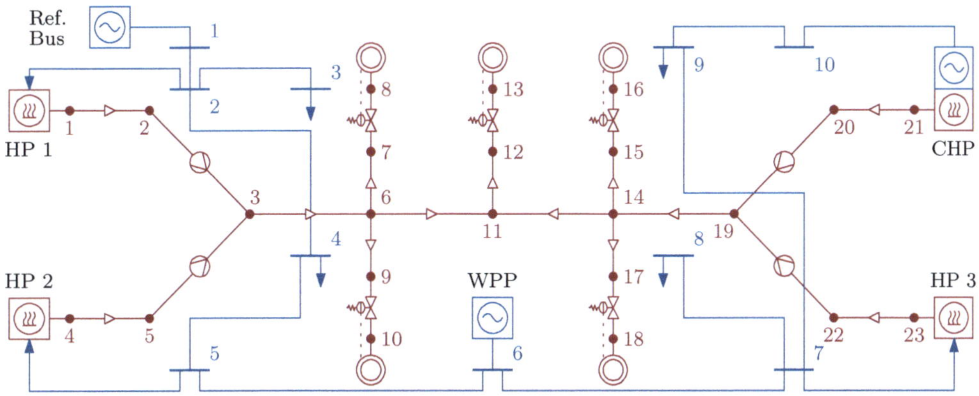

Figure 4.4: Schematic diagram of the EPN (shown in blue) and the supply network of the DHN (drawn in red color); the symmetric DHN return network shown in Figure 4.5 is omitted here for simplicity.

no power exchange with the transmission system; this is facilitated by placing a BESS at the reference bus, which can be considered as a challenging EPN operating condition. The heat pumps are modeled as flexible consumers in the EPN and flexible producers in the DHN.

District Heating Network The DHN supplies all consumers with heat, injected by the HPs and the CHP, as shown in Figure 4.5. The FNP at Node 13 is a large-scale consumer (e.g. a secondary network) while the other FNPs are small-scale consumers. This DHN has a meshed structure due to the presence of the supply and return network; the flow directions are determined in advance through the operational states of the pumps and the set point ranges of the valves. The effects of check valves in consumer and producer facilities prevent flow reversals, which are taken into account in (3.14). The used pipeline model is a special case of the model presented in Section 2.3.3, as VMFDs can be neglected, see [MTK$^+$22] for details. The pumps are located next to the producer facilities, i.e., between Nodes 2 and 3, and the DPRs are found in the consumer lines, such as the one between Nodes 7 and 8, which ensures that the pressure difference between Nodes 8 and 34 remains constant by varying the differential pressure between Nodes 7 and 8. The temperature limits which ensure secure[8] and reliable operation of the supply network are set to 80 °C and 130 °C, as in [NTJK18], and the minimum output temperature of all producers is assumed to be 95 °C, for efficiency reasons. An ambient temperature of 10 °C is considered, as this is a typical autumn temperature in Germany. The hydraulic network parameters, such as the roughness, diameters, and lengths of the pipelines, are taken from [Liu13]. The hydraulic model variables β and μ in (2.21) are modeled as parameters based initial mass flow values, see [MTK$^+$22].

Grid Participants The system base power profile and associated peak demands are defined by assuming that consumers in the power system and DHN are not participating in the TCS auction market. It is further assumed that consumers are capable of reducing their power consumption up to 30 % at any point in time. This allows to properly demonstrate the performance of the proposed framework and the models; in practice, this flexibility would be

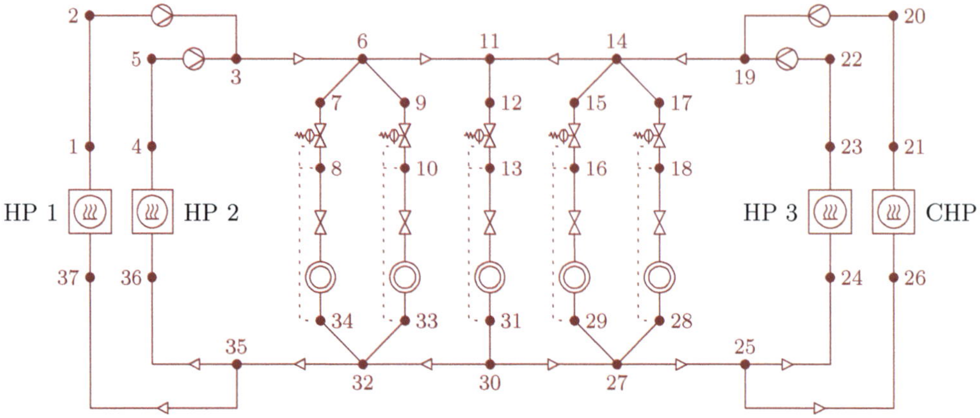

Figure 4.5: Schematic diagram of the meshed district heating network with supply and return network.

Table 4.2: Cost and benefit parameters of flexible grid participants in general monetary units ¤ for Scenario 1.

		c_i in ¤/MW
EPN	BESS	25
	CHP	30
	Heat pump	12
	Flexible Consumer	9
DHN	CHP	20
	Heat pump	6
	Flexible Consumer	7

determined by their EMSSAs by first calculating the predicted minimum power demand for the next hours and the expected usable amount of energy, as outlaid in Section 3.1. The power profiles of consumers are based on the electric load profiles of [Gmb18], and the heat load profiles of [YS05a], scaling, time-shifting, and overlaying them with a noise profile to represent different consumers. The power profile of the WPP was taken from [Bun19a], and was scaled down and shifted. The coupling factor of the CHP was set to 1.42, and the coupling factors of the heat pumps were set to 4, based on [NTJK18].

The cost and benefit parameters c_i of producers and consumers are given in Table 4.2. The same values for c_i are used here for all time steps k in order to facilitate the analysis of the simulation outcome; however, results with varying cost parameters are also tested and discussed in [MTK+22]. It is important to mention that the power injection as well as power consumption of the reference bus/BESS results in income for the battery in the form of an ancillary service, which can be interpreted as an incentive for storage systems.

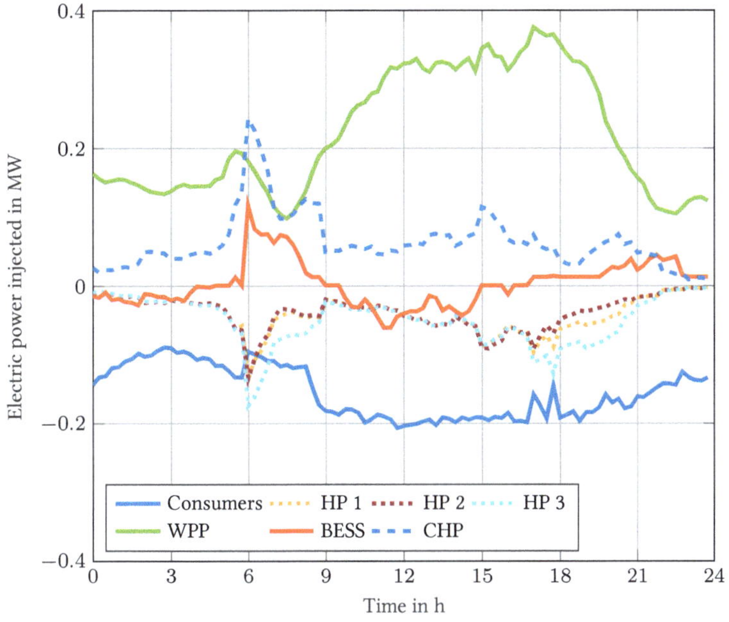

Figure 4.6: Injected power by all EPN participants for Scenario 1.

Results and Analysis

The prediction horizon of the ISOEMS is set to 16 time steps, of 15 min each, which is the interval used in the German intraday market, and the total simulation horizon is 24 h. Several forms can be considered using the proposed approach to provide additional flexibility to the power system. In particular and as demonstrated next, one option is to use DSM of electric and heat consumers to bridge a period with high power price. This leads to a cost advantage for the consumers, allowing to minimize the power injected by more expensive nonrenewable energy sources. Another option is the use of HPs during a period of high penetration with RESs to transfer energy to the DHN.

The electric power injected by EPN participants, calculated using the proposed approach, is shown in Figure 4.6, where the power injected by consumers is shown in aggregated form. A more detailed view of the consumer power is illustrated in Figure 4.7 where the maximum and actual aggregated power of all electrical and heat consumers is shown. To understand the temporal progression of the electric consumer power, a comparison with the power of the reference bus and the WPP of Figure 4.6 is helpful. Thus, during a period with a large ratio between the infeed of the WPP and the aggregated demand, the power of the electrical consumers is maximized in order to minimize the power transferred by the reference bus; this is the case between 0 h and 5.25 h as well as between 9.00 h and 16.75 h. The rest of the time, the consumed power is reduced to its assumed minimum of 30 % of the maximum power demand. This allows minimizing the power injected by the reference bus and the CHP in the periods where the injection of the WPP cannot provide the entire power demand. To illustrate

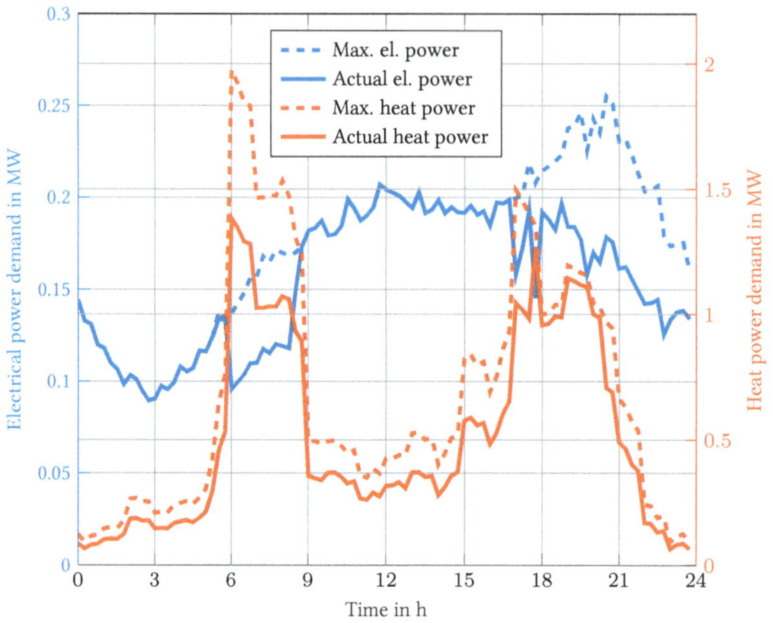

Figure 4.7: Aggregated electric and heat power demand of all consumers for Scenario 1.

the behavior of the HPs, the output temperatures at Node 1, 4, and 23, in Figure 4.8 can be used in addition to Figure 4.6. In contrast to the consumers, the power of a HP is coupled with the output temperature through the temperature ramping constraint (3.25); this causes a strong correlation to the current mass flow flowing through the HP, as per (2.59). Thus, during the period between 8.75 h and 15.5 h, the HPs consume their maximum power; a further increase of the heat power output is not feasible due to the nodal temperature constraints (3.16), as depicted in Figure 4.8. During the periods between 7.75 h and 8.75 h and 22.25 h and 23.5 h, the power consumption of the HPs is at its minimum. The sharp changes in the temperature are due to the varying wind power infeed, as may be noted in Figure 4.6, and the varying electric and heat power demand, as shown in Figure 4.7.

The effective flexibility provision by the DHN can be illustrated in Figure 4.9, where the balance between the aggregated power injection and consumption in the DHN can be observed. The average node temperature and the power infeed of the WPP is also shown to highlight the correlation to the heat power balance. This Figure illustrates the energy that can be stored in the DHN; thus, the DHN provides flexibility to the EPN because excess power from it can be transferred to the DHN, supplying heat consumers with heat power. In Figure 4.9, it can also be observed, that the TCS achieves a very efficient form of DHN operation with low losses, as temperatures in the DHN are continually reduced, during periods when low cost power infeed of the WPP is sparely available. The described effect is observed from the profile of the WPP power infeed and the decreasing average node temperature, with a maximum decline of 30 °C, between 17.5 h and 24 h. This is particularly relevant, as reducing the temperatures has the largest impact on reducing heat losses in a DHN [ÇYÇ04].

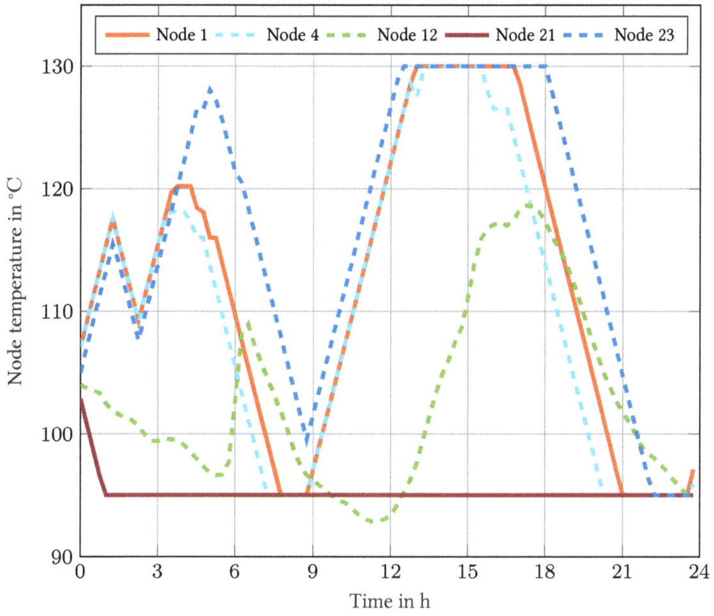

Figure 4.8: Node temperatures in the DHN for Scenario 1.

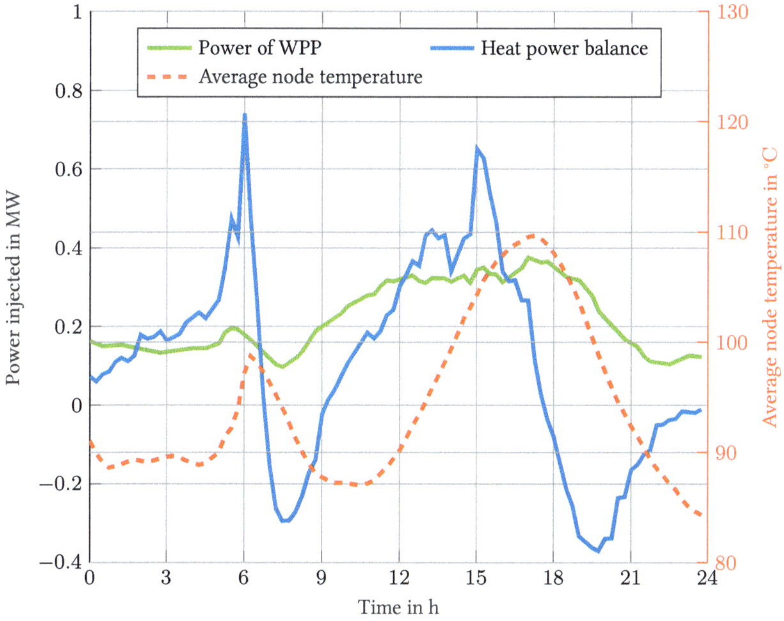

Figure 4.9: Power balance and average node temperature in the DHN and injected power of the WPP in the EPN for Scenario 1.

By design, the TCS inherently minimizes the losses in the CEPDHN, which results from two important aspects. The first is that the prediction horizon of the underlying NMPC control mechanism is limited; thereby, the ISOEMS is continually incentivized to reduce the temperatures in the DHN to the lower operational limits toward the end of the prediction horizon, in order to maximize the welfare in (3.5) by reducing the allocated power infeed into the DHN. Therefore, the TCS always uses the stored energy in the DHN pipelines as soon as possible, for welfare maximization. The second is the simple fact that lower network losses in the entire CEPDHN need less power infeed, which also inherently incentivizes the ISOEMS to reduce losses to maximize (3.5). These aspects show how market and control mechanisms of the proposed TC approach optimally collaborate to achieve technical efficiency and economic optimality.

The optimization was implemented in GAMS using the IPOPT solver [NW06]. After the initialization, each simulated time step, which includes the 16 prediction horizon time steps, took less than 2 min 28.97 s of CPU time. The entire simulation of all 96 time steps for this scenario took 76 min 35.20 s on the same hardware as Case Study I, see Section 4.1.1. This enables computation in real time even without exploiting the possibilities of distributed parallel optimization or faster workstations.

4.2.2 Scenario 2: Price Signals and Independent versus Coupled Operation of EPN and DHN

Description

The second scenario depicts a cold winter day with an ambient temperature of T_a of $-10\,°C$. The BESS at the reference bus in Figure 4.4 is replaced by a Photovoltaic (PV) plant, and the profile of the WPP corresponds to data for the simulated day; both RES profiles where taken from [Bun19a] and scaled down to match the demand profiles. Furthermore, RESs are treated as all other flexible producers and can thus be curtailed, which results in all auction bids coming from fully flexible FNPs, that can reduce their demand and supply by 100 %. Inflexible demand, from INPs, is assumed to be allocated in earlier cleared markets or long-term contracts. To show the operation for different hydraulic conditions, the operating points of pumps can be set variably by the TCS in this scenario. Thereby, the TCS receives additional degrees of freedom. Finally, the assumed bid and offer prices are shown in Table 4.3.

Results and Analysis - Price Signals

The resulting heat and power injection and demand, are shown in Figure 4.10, where it can be seen that HPs are preferred over other flexible electric consumers by the TCS. This is particularly obvious between 5.5 h and 8.5 h when the heat demand is large, and the RESs infeed is not sufficient to meet the entire electricity demand. Furthermore, note that the TCS utilizes HP 3 the most, as it is the highest contributor to the social welfare, while the most expensive producer, the CHP, is only run between 5.25 h and 8.5 h to supply the peak demand.

Table 4.3: Cost and benefit parameters of FNPs in general monetary units ¤ for Scenario 2.

		c_i in ¤/MW
EPN	CHP	10
	PV	7
	WPP	4
	HP 1	12
	HP 2	9
	HP 3	9
	Flexible Consumer Bus 3	12
	Flexible Consumer Bus 4	11.5
	Flexible Consumer Bus 8	11
	Flexible Consumer Bus 9	10
DHN	CHP	7
	HP 1	4.2
	HP 2	3.5
	HP 3	3.2
	Flexible Consumer Node 8	11
	Flexible Consumer Node 10	10.8
	Flexible Consumer Node 13	10.5
	Flexible Consumer Node 16	10.2
	Flexible Consumer Node 18	10.9

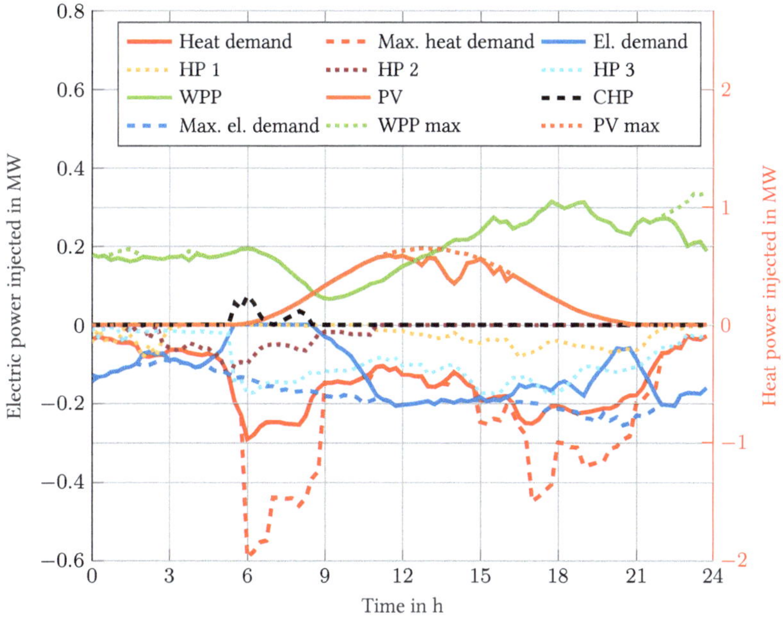

Figure 4.10: Injected power by all EPN participants and all DHN consumers for Scenario 2. The power injection of the CHP plant and the HPs corresponds to electric power.

The LMPs for this scenario are calculated from the optimization model, and the UMPDs are based on the HPA, described in Section 3.2.3. The UMPDs of the EPN and the DHN are calculated separately, based on the intersection of the dispatched offer and bid curves, which are established by sorting the dispatched bids and offers by their respective prices, as outlaid in Section 3.2.3. The resulting LMPs and UMPDs[115] are shown in Figure 4.11, respectively. Note, that in Figure 4.11 the LMPs show large variations, which may be attributed to some extent to the dynamics of temperature propagation throughout the network and the coupling component between the EPN and the DHN [DLS+19].

The UMPD of the DHN in Figure 4.12 shows a strong correlation with the heat power demand of the DHN, which is a logical result of the power sources accessed by the Merit Order. Also, a correlation between the RESs infeed, the total of the WPP and the PV power supply, and the UMPD of the EPN can be seen. Thereby, showing lower mean UMPDs towards the mid of the day during the time of high power supply, of low price offers of the RESs. A comparison of the UMPDs with the marginal bids and offers prices in Table 4.3 yields that the marginal units in the EPN price zone are the WPP, the HP 2 and 3, as well as the PV plant. In contrary, all FNPs in the DHN except the two consumers at Node 10 and 18, with the highest benefits determine the UMPDs at a certain time step. Despite the well known advantages of LMPs in the context of market-based EPN operation such a pricing approach does not seem to work in this case since two different LMPs are obtained for one FNP, as described in Section 3.2.3. Therefore, the UMPD-based pricing seems more appropriate for the application in the presented TCS. This pricing approach incentivizes the FNPs to bid with their true marginal costs in the absence of network congestions [MGR+21].

Results and Analysis - Independent versus Coupled Operation of EPN and DHN

In order to examine the performance of independent operation of the EPN and the DHN, in contrast to their coupled operation, five different operating cases were considered as illustrated in Table 4.4. For Case 3 and 4 one network was optimized first, fixing the obtained power of the energy converters to optimize the second network sequentially. These resemble the operation where either the heat or the electric power market would be cleared first, without any information exchange between the markets. In both cases, the optimization of the second network failed to converge in several time steps, as noted in Table 4.4. Precisely, for Case 3, the DHN optimization problem is terminated by the solver in the first optimization due to the set iteration limit at 8000 iterations of the IPOPT solver. The calculation time needed therefore is beyond 27 min. However, still multiple constraint violations exist, including the equality constraints fixating the heat power infeed of the energy converters to the values obtained from the EPN optimization[116]. In Case 4, the EPN optimization fails the first time during the optimization performed for the 2 h time step. Thereby the infeasibility occurs at the real power flow constraint (2.15) of Bus 2 for the 16 time step, which is the last on the prediction horizon and represents the time step at 6 h. Note, that this is the time step with the maximum heat

[115] The loss costs are set to 0 ¤/MW for the EPN and the DHN.

[116] For comparison, running the identical DHN optimization without the equality constraints, fixating the heat power infeed of the energy converters to the values obtained from the EPN optimization, leads to conversion within 226 iterations.

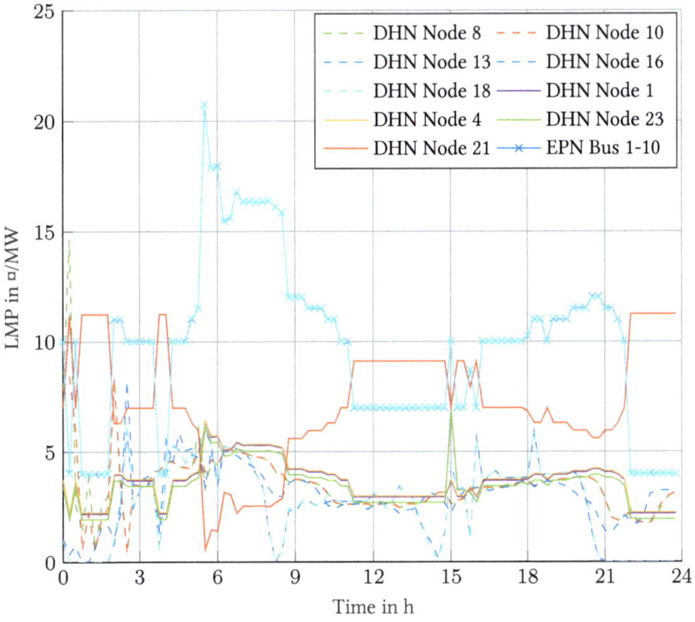

Figure 4.11: LMPs of EPN and DHN for Scenario 2.

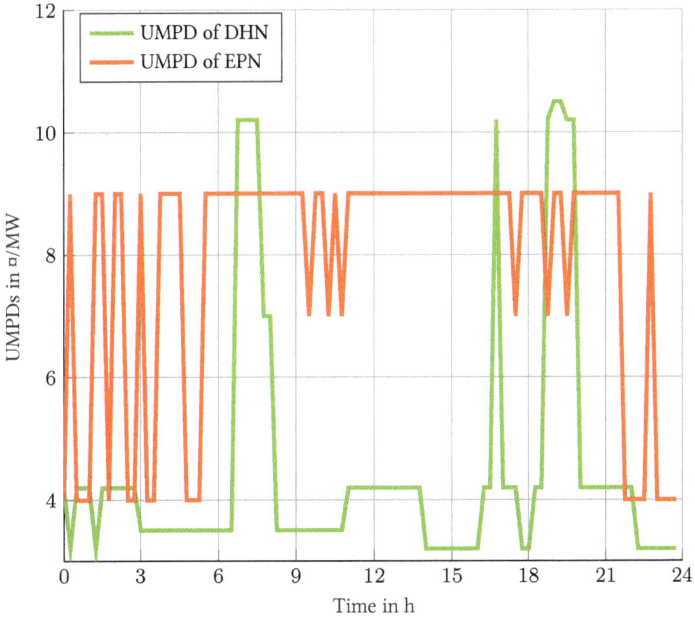

Figure 4.12: UMPDs of EPN and DHN for Scenario 2.

Table 4.4: Accumulated 24 h social welfare in various cases.

Case		Social welfare in ¤
1.	EPN optimized operation, no DHN	187.39
2.	DHN optimized operation, no EPN	321.77
3.	EPN, DHN optimized in sequence	No Feasible Solution
4.	DHN, EPN optimized in sequence	No Feasible Solution
5.	EPN and DHN joint optimization	443.70 (EPN = 138.06; DHN = 305.65)

power demand, see Figure 4.10. In this case, the first dispatched heat demand is so high, that the resulting EPN load created by the HPs can no longer be provided by the EPN sources. Thus, the optimization fails. Thus, a secure[8] network operation can not be provided in both energy networks in either Case 3 or Case 4.

A comparison of the Cases 1, 2 and 5 for their 24 h social welfare, given in Table 4.4, shows that the joint optimization of the EPN and DHN yields the highest social welfare, whereas, if the EPN or the DHN are optimized independently, without considering the presence of the other network, the social welfare is significantly reduced. It should be noted, that the welfare obtained from the DHN is larger as the same accumulated from the EPN operation, as the heat demand is respectively higher, see Figure 4.10[117].

4.3 Case Study III: DHN Operation with Variable Mass Flow Directions and Thermal Energy Storage Systems

Within this case study optimal market-based DHN operation of the TCS is performed, presenting the effects of VMFDs and the functioning of a large TESS.

4.3.1 Description

The regarded DHN is similar to the network used in Case Study II which is extended here by the TESS, including respective edges for the storage valve and pump, see Figure 4.13. The large scale TESS is directly connected to the DHN and run by the network operator, as described in Remark 3.10. The TESS has a maximum storage capacity of 20 MWh and is initialized almost empty, with 0.1 MWh. The losses of the TESS are assumed to be constant with a value of $l = 1$ kWh. Further, VMFDs are enabled through adapted valve and pump set point ranges. Based on this, variable mass flow directions can occur on the depicted edges, see Figure 4.13. The marginal bid and offer prices of all FNPs are given in Table 4.5. Note, that these vary over time as stated. The bids and offers of the TESS are set to $c_{n,k} = 0$ for all time steps k, thereby operating it in a way that serves the good of all other market participants. Further, the

117 The dashed red and blue lines show the maximum heat and power demand.

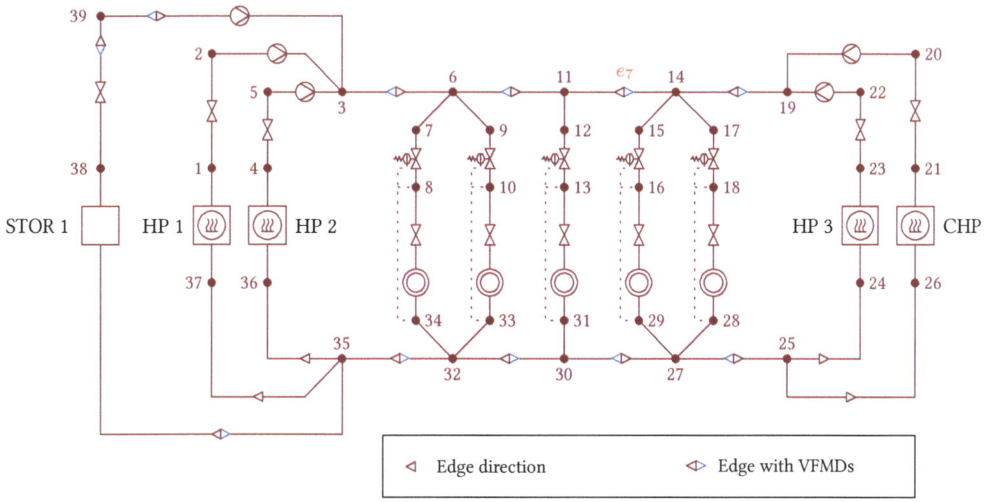

Figure 4.13: DHN with supply and return network with occurring VMFDs and TESS connected.

factors defining the approximated rectangular function (3.7), used to bounds the mass flows away from zero, are set to $w_{e,k}^{\text{rect,w}} = 1\text{e}^{-2}$ and $w_{e,k}^{\text{rect,h}} = 5\text{e}^{-2}$. The prediction horizon was set to 10 time steps. The optimization of all simulation time steps were calculated within no more then 89 s.

4.3.2 Results and Analysis

The aggregated heat consumption of all consumers and the heat power injection of all producers and the TESS are shown in Figure 4.14. By comparing the power infeed of the different producers and the offer costs in Table 4.5 it can be observed that after the initialization the aggregated consumed power of all consumers stays fairly constant. Besides the ISOEMS maximizes the welfare by prioritizing the producers offering heat at lower costs. These are HP 3 and the CHP toward the beginning and the end of the simulation and HP 1 and HP 2 between simulation time steps k_{16} and k_{26}. This leads to VMFDs in the DHN which are exemplary discussed here for the pipeline on Edge 7, marked as e_7 in Figure 4.13. The respective mass flow is plotted in Figure 4.15. As HP 3 and the CHP are the cheapest energy sources at the beginning and the end of the simulation during these time steps heated water flows from these producers over Edge 7 towards the consumer connected at Node 13. In contrast, for $k_{13} < k < k_{28}$ the flow direction is reversed as the producers on the left side of the DHN provide the main power demand. Note, that the storage is strongly charged during the time interval between simulation time steps k_{12} and k_{14}, when the ISOEMS anticipates, that heat power will become more expensive in the next time steps, based on the offer prices. Similarly, the storage is strongly discharged in the following period until k_{27} when the offer prices for the upcoming time steps, sent to the ISOEMS show falling offer prices in the next future. Thereby the adapted TESS operation enlarges the overall welfare.

Table 4.5: Cost and benefit parameters of FNPs in general monetary units ¤ for different simulation time steps k.

FNP	c_i in ¤/MW for $k < k_{16}$	c_i in ¤/MW for $k_{16} \leq k < k_{27}$	c_i in ¤/MW for $k_{27} \leq k$
HP 1	12	12	12
HP 2	13	13	13
HP 3	10	30	5
CHP	11	30	7
TESS	0	0	0
Flexible Consumer Node 8	21	21	21
Flexible Consumer Node 10	22	22	22
Flexible Consumer Node 13	23	23	23
Flexible Consumer Node 16	24	24	24
Flexible Consumer Node 18	25	25	25

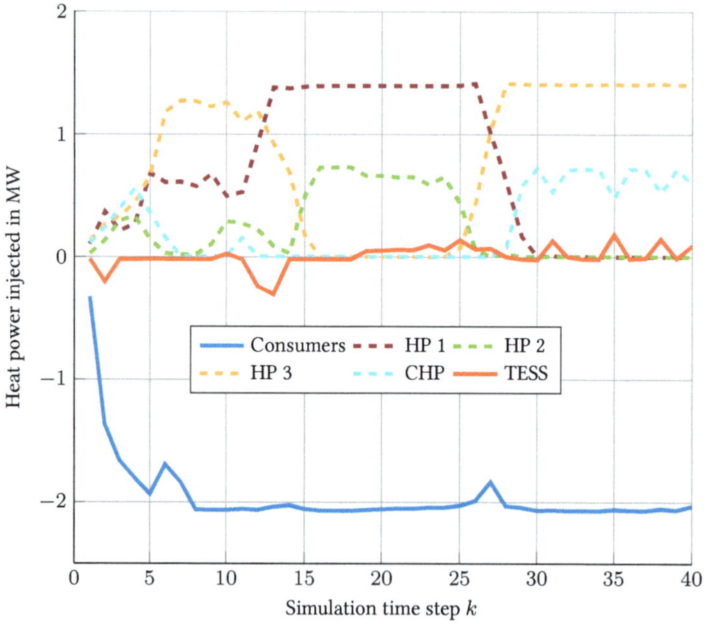

Figure 4.14: Injected power by all DHN participants for multiple simulation time steps.

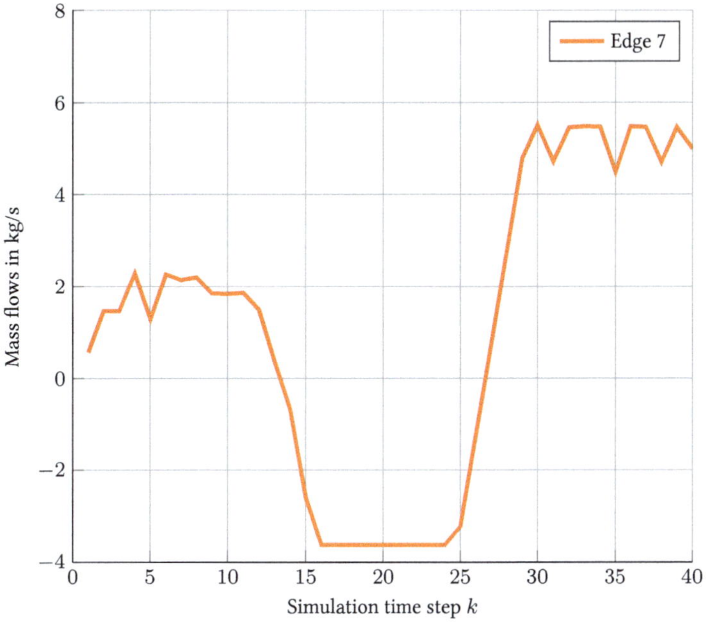

Figure 4.15: Mass flows on Edge 7 for multiple simulation time steps.

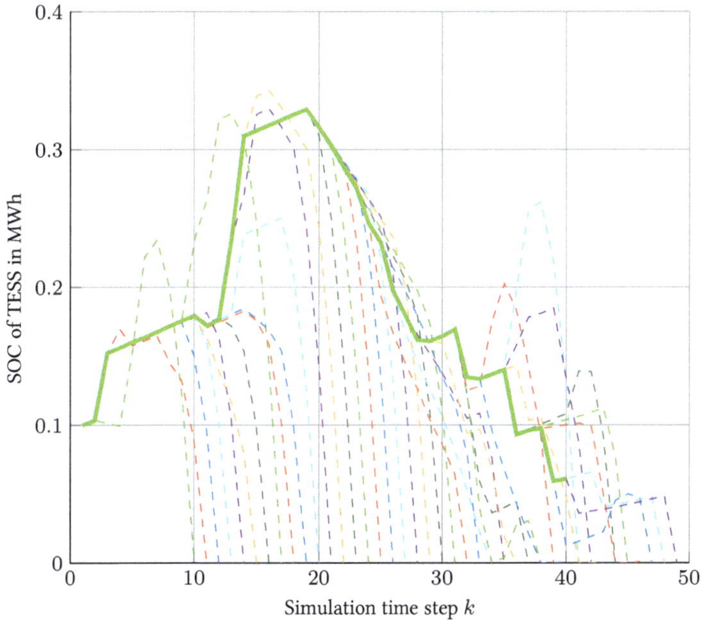

Figure 4.16: State of charge of the TESS over for multiple simulation time steps and calculated SOCs by the ISOEMS over the prediction horizon at these simulation time steps.

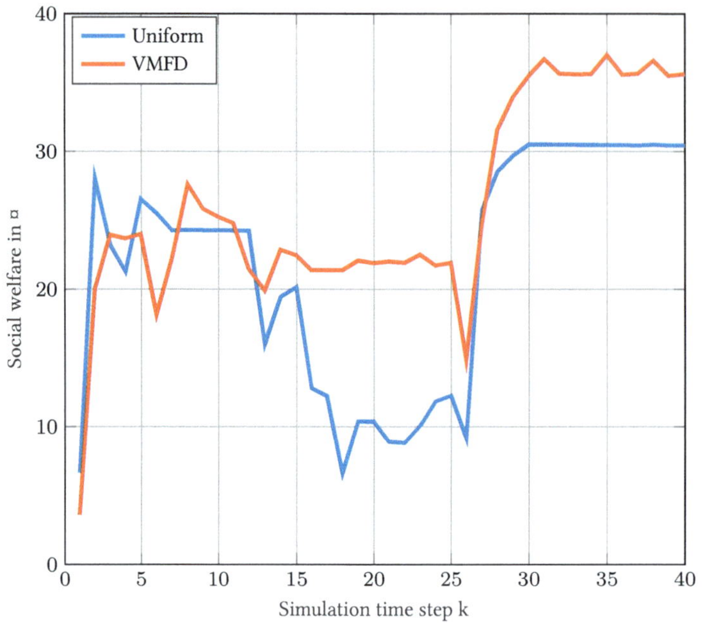

Figure 4.17: Social welfare obtained in the uniform and VMFDs case for multiple simulation time steps.

In order to compare the VMFD case with a form of network operation with fixed flow directions, the same Scenario was also simulated with fixed (positive) flow directions on all edges, except the edges of the TESS between Node 3 and Node 35, see Figure 4.13. The resulting social welfare for both cases, with and without VMFDs, is depicted in Figure 4.17. Therefrom it can be seen, that after the initialization, the difference in the social welfare increases, with increasing offer price deviation between the producers on the left and right side of the DHN. As can be seen in Table 4.5 these offer price deviations are maximal between $k_{16} \leq k < k_{27}$ with up to 18 ¤/MW and remain high with 8 ¤/MW for the rest of the simulated time steps $k_{27} \leq k$. This demonstrates, how changing the flow directions in the respective edges, enables the ISOEMS to provide the consumers with energy from the heat sources with lower offer prices. The accumulated social welfare over the regarded 40 simulation time steps adds up to 863.5 ¤ in the uniform flow direction case and up to 1042.4 ¤ in the case of VMFDs. This shows that using a form of operation enabling VMFDs raises the social welfare by 20.72 % in the provided scenario.

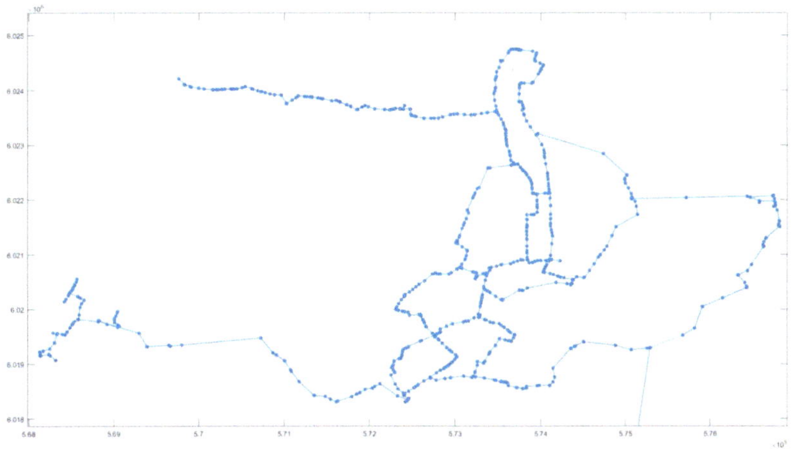

Figure 4.18: Main network of DHN based on raw data.

4.4 Case Study IV: Calculation Times in Large Coupled Electric Power and District Heating Networks

The aim of the following case study was to test the calculational burden of the rolling horizon approach based on the CEPDHN model of a large scale network. As the computational burden of the DHN represents the central challenge, the approach was tested on one of the fifteen biggest out of 1169 DHNs in Germany [SS12], the network of the municipal utilities of a city in northern Germany.

4.4.1 Description

District Heating Network

The main network of this DHN consists of 2864 nodes and 2894 edges and is shown in Figure 4.18.

In order to reduce the amount of optimization variables for the optimal operation approach the network model was aggregated first [BHK+02]. For that matter, serial pipelines are merged into a single component, if they are the only components connected to the respective DHN node in between. This procedure reduces the network graph to 300 nodes and 367 edges. Further, possible future producers in the year 2030 were determined based on a design problem minimizing operation costs taking into account the fees stemming from CO_2 emissions. The results of these calculations proposed the installation of 11 additional HPs with 4 MW thermal each. The used connection of these HPs to the supply network is indicated in Figure 4.19.

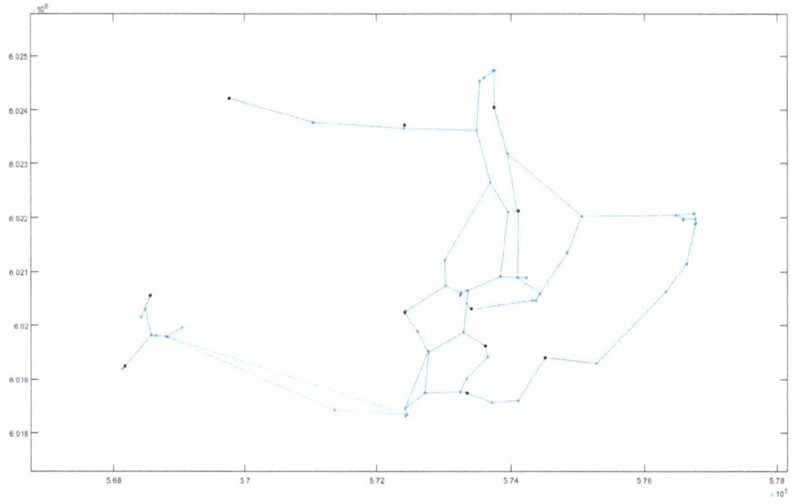

Figure 4.19: Aggregated supply network of the DHN with connected HPs marked in black.

The thermal resistance, roughness, length, and diameters of the pipelines were set to typical values of $0.3\,\mathrm{m\,K/W} \leq R^{\mathrm{therm}} \leq 0.6\,\mathrm{m\,K/W}$, $r = 0.01$ mm, $L = 50\,\mathrm{m}$[118], and $d = 300\,\mathrm{mm}$, as given in [NTJK20, p. 64], [Liu13, p. 144]. The amount of positions reserved within a pipeline to model water masses $|\mathbb{S}_\zeta|$ was set to 20. The producers supplying heat to the DHN are CHPs, EBs, HPs, heating plants and a waste incineration plant. The maximal total heat demand is around 350 MW. The 100 biggest consumers account for 33 % of the heat demand [MB17]. It is assumed, that these largest consumers are used for DSM with the ability to time shift their demand by 3 h.

Electric Power Network

The graph of the 110 kV EPN is shown in Figure 4.20. All assumptions on the present producers and consumers[119], including values of variable power demand at every bus, are documented in [Vie19]. Electric power supply comes from the super ordinate 220 kV network connected at Bus 8 and 9, CHPs and a waste incineration plant. The power demand includes the electric demand of the EB and the newly installed HPs which are connected at the Buses 1, 2, 4 and 5. The volume of electric energy sold by the municipal utilities per year is around 733 GWh [Sta].

118 The set value is understood as a mean value stemming from aggregated pipeline lengths. Therefore, the lengths of the new pipelines are calculated by the sum of the aggregated serial pipes [BHK+02].

119 Except the HPs.

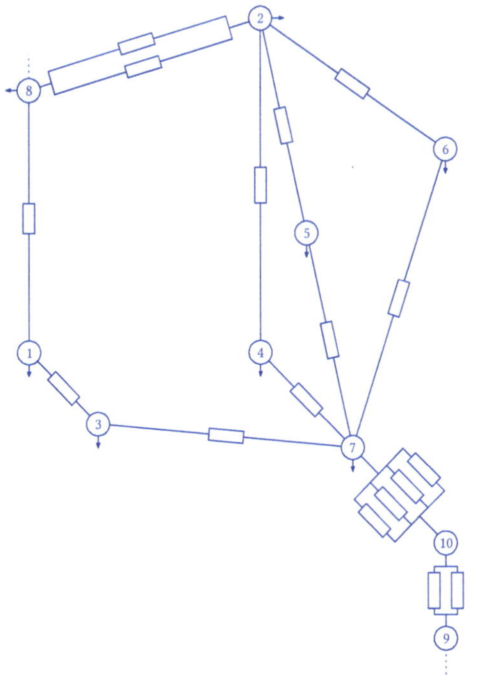

Figure 4.20: Topology of the 110 kV electric power network [Vie19].

Implementation

The pre-processing of the raw data was implemented in MATLAB. The optimization problems solved for the rolling horizon approach were implemented in GAMS. The used optimization problem was slightly altered to fit more into existing network operation forms of network operators. The parameters defining the soft constraints f^{soft} within the objective function (3.5), bounding the mass flows away from zero, are set to $w_{e,k}^{\text{rect,w}} = 1\text{e}^{-3}$ and $w_{e,k}^{\text{rect,h}} = 2.5\text{e}^{-1}$. Further, the economic objective f^{eco} in (3.5) was set to minimize the operation cost of the DSO within the network, including resulting costs for the network operator when applying DSM. Thus, the idea behind this approach is the design of an EMS for a DSO which is not unbundled, in order to minimize its operation costs. Note, that even though the objective function is therefore slightly altered, the basic functionality of the approach described in Figure 3.7 stays identical, except that the cost parameters needed in f^{eco} are set and known by the DSO and Step 3 is omitted. Thus, also the calculation time needed to solve the optimization problem in Step 2, is in a similar range as for a central ISOEMS developed within this work, as it uses the same CEPDHN model given in Chapter 2 and operational constraints outlaid in Section 3.3.1.

4.4.2 Results and Analysis

The obtained calculation times for different prediction horizons are shown in Figure 4.21 for three subsequent optimizations on the prediction horizon each. Note, that for a prediction horizon of five time steps $n^{\text{ck}} = |\mathbb{S}_k| = 5$, the solver has returned an error message in the first[120] and second[121] optimizations, thus these calculation times are not provided in the graph. Here, the solver tries to allocate more memory multiple times before in the first optimization before aborting the solution procedure. The results show, that real time capability of the approach on standard working stations with of the shelf solvers is not given for large DHNs . This is depicted in the 4.21, through the surpassing of the time step interval $\Delta k = 15\,\text{min}$ used between two optimizations by multiple calculation times of different optimizations. It should also be taken into account, that larger prediction horizons are desired to enable the energy management system to better anticipate future operation conditions, in order to minimize losses and minimize operation costs, in the application shown within this case study. Note, that for a $1\,\text{h}$ prediction horizon, in certain low flow conditions the supply of the critical consumer will not be represented within the DHN model used by the EMS. This shows, that larger prediction horizons would be favored. Note, that even though real time capability is not reached here, the necessary reduction of computation times seams possible with more resources, as necessary reduction factors are assumed to be in the single digit range even for larger prediction horizons. Thus in order to enable real time capability, further measures need to be taken. These could comprise faster work stations, better initial values $\boldsymbol{x}^0$ obtained from precalculations with simplified models [Trö99], and the elimination of variables [NW06, p. 426]. Besides the aforementioned, using parallelization techniques, as the OCD approach,

[120] IPOPT message: Problem with integer stack size.
[121] IPOPT message: Restoration failed.

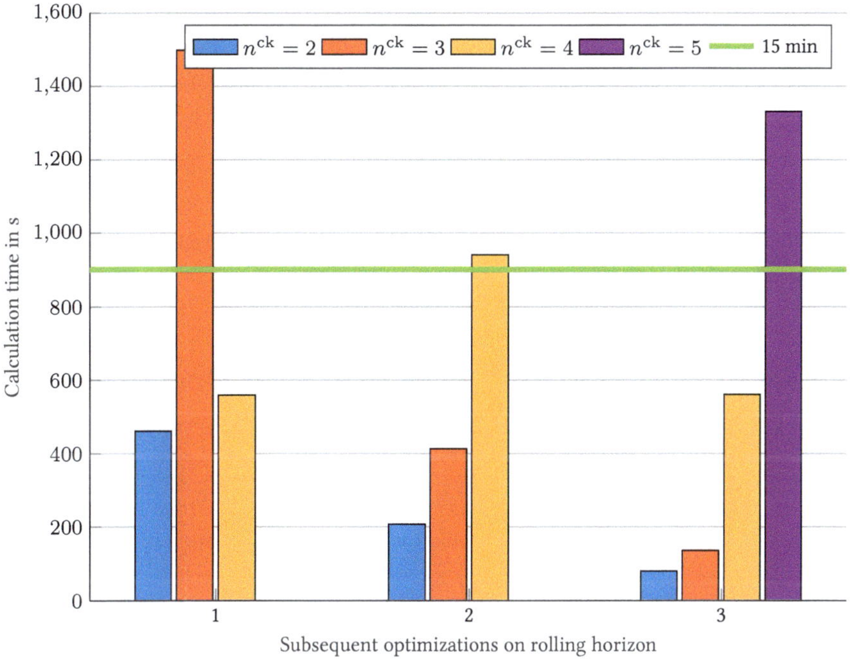

Figure 4.21: Calculation times needed for different prediction horizons in subsequent optimizations on the rolling horizon.

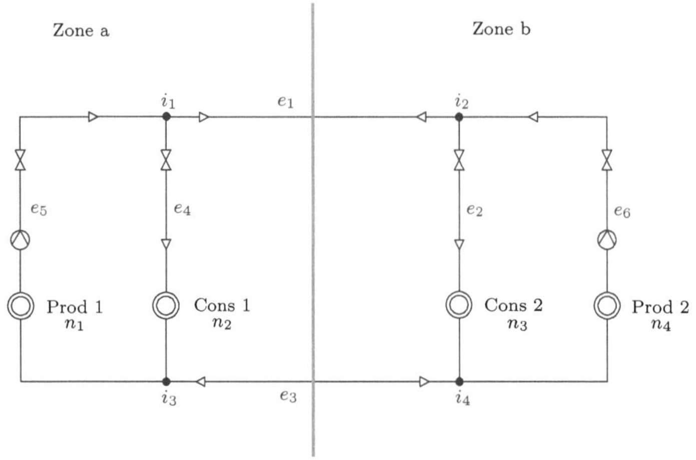

Figure 4.22: Regarded DHN with two producers and consumers each decomposed by the orange marked operational border into Zone a and Zone b.

could help to reduce the calculation times. In this context, the basic principle of applying OCD to a DHN is demonstrated in the following Case Study V.

4.5 Case Study V: Optimality Condition Decomposition for Operation of District Heating Networks

Within this case study the applicability range of distributed optimal operation of DHNs based on OCD is analyzed. The approach was tested on two DHNs presented below.

4.5.1 Description

Four Node Network

The regarded DHN is shown in Figure 4.22. Note that, this DHN has already been discussed extensively in Section 3.3.2, where the approach has been presented. The case study was implemented using the aggregated component model form described in Remark 3.19. In doing so, the network can be described with only four network nodes i and six network edges e. As explained in Remark 3.20 the border edge directions are set to direct away from the border nodes i in the supply network and towards both border nodes in the return network, within the regarded DHN node edge incidence matrix $\boldsymbol{A}^{\mathrm{dhn}}$. The optimization is initialized with a flat start, thus setting $\boldsymbol{x} = \mathbf{0}$. Mass flows have been limited to $0\,\mathrm{kg/s} \leq \dot{m}_e \leq 2\,\mathrm{kg/s}$, permissible outlet, inlet and nodal temperatures are defined as $70\,°\mathrm{C} \leq T \leq 100\,°\mathrm{C}$. The box constraints of the injected power infeed of the producers on Edges e_5 and e_7 are $0\,\mathrm{kW} \leq \Phi \leq 40\,\mathrm{kW}$, while consumers on Edges e_2 and e_4 are limited to $-40\,\mathrm{kW} \leq \Phi \leq 0\,\mathrm{kW}$. The predefined pressure coefficient $\mu_{e,k}^{\text{pre}}$ of the pipelines was set to 1 and the predefined pressure value p^{pre} used in (3.53) was set to 0.2. The marginal bid prices c_n of the FNPs n_1, n_2, n_3, and n_4 on Edges e_5, e_4, e_2 and e_7 are set to 1, 10, 20, and 20. Note, that n_1 and n_4 are producers while n_2 and n_3 are consumers, as can be seen from Figure 4.22. The approach to guard against singularity of the approximated KKT matrix was used, as explained in Remark 3.17, with the parameter $\delta = 1e^{-6}$.

Eight Node Network

A second network was tested, which represents an extension of the Four Node Network described in the previous Section. In this network a pipeline is added before and after the two producers shown in Figure 4.22. In order to connect these newly inserted pipelines with the producers, also a node is added between each pipeline and every producer. Thus, in total the network shown in Figure 4.22 is extended by four pipes and four nodes. The entire set of parameters used is provided in [MIJSH22].

4.5.2 Results and Analysis

Four Node Network

The results obtained from the approach presented in Section 3.3.2 for two ISOEMSs is compared here against the results obtained from a single central ISOEMS. The solution of the central ISOEMS case is shown in dashed lines within the plots of Figure 4.23. As can be observed from these graphs the mass flows, the nodal pressures, the node temperatures and the power infeed or demand of the FNPs converges to the same values for the distributed ISOEMS case[122]. The maximum deviation of the plotted values between the central and the distributed solution is obtained for the node temperature at Node i_3 with $-0.157\,\text{mK}$. The central solution, implemented in GAMS with the IPOPT solver needs 29 iterations to converge to the standard solver tolerances, while the distributed approach achieves this accuracy after 100 iterations. This need for more iterations is expected, as the search direction is calculated less accurate in the distributed case as in the central implementation, which leads to a need of additional iterations. This finding accords with the results in [CNP02, NPC03]. As the calculation times of the single iterations could be reduced in the distributed case, for lager networks and thus optimization problems, the distributed case may reduce the calculation times as demonstrated in [CNP02] for EPNs, also see Remark 3.18. The primal infeasibility[123] $\|\boldsymbol{h}(\boldsymbol{x})\|_\infty$ of the subproblems per iteration ν is shown in Figure 4.24 on a logarithmic scale. Therefrom it can be seen, that both subproblems fulfill the set constraints with a maximum constraint violation below $1.1e^{-4}$ after 100 iterations. Note, that similar results were also obtained for different sets of parameters within this network, see [MIJSH22].

122 Note, that the mass flows of Edge 1 and 2 are nearly the same $\dot{m}_1 \approx \dot{m}_2$, as the flow through Edge 7 is $\dot{m}_6 \approx 0$, also $T_1 \approx T_2$, thus these values are hard to distinguish.

123 Based on the IPOPT NLP problem formulation [Wäc09, p. 7].

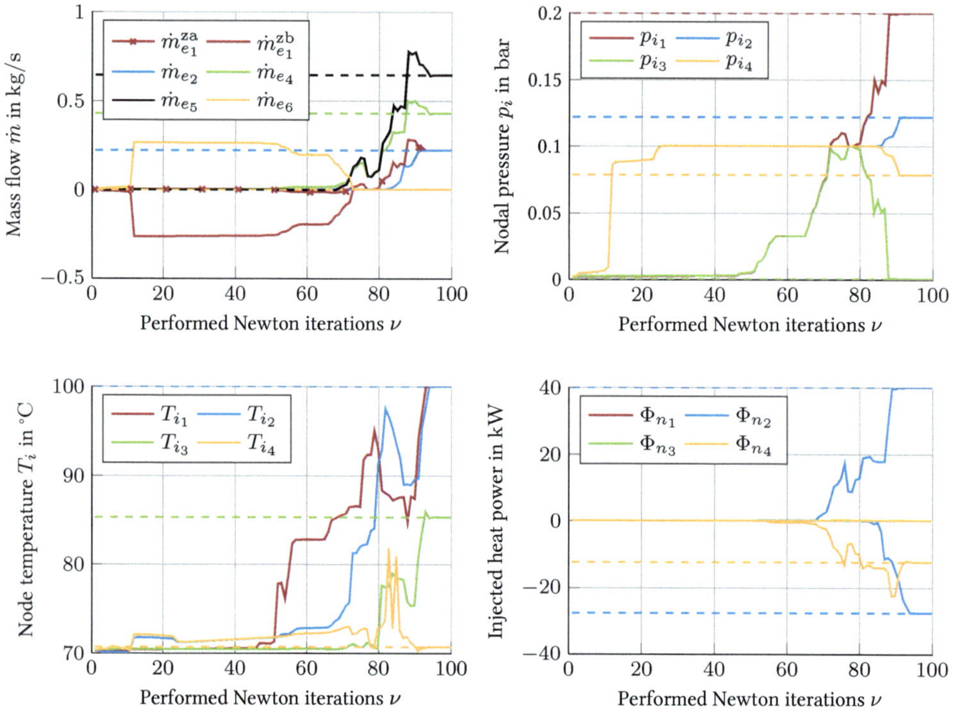

Figure 4.23: Four most relevant variables plotted for the performed OCD iterations. The dashed lines show the solution of the central optimization.

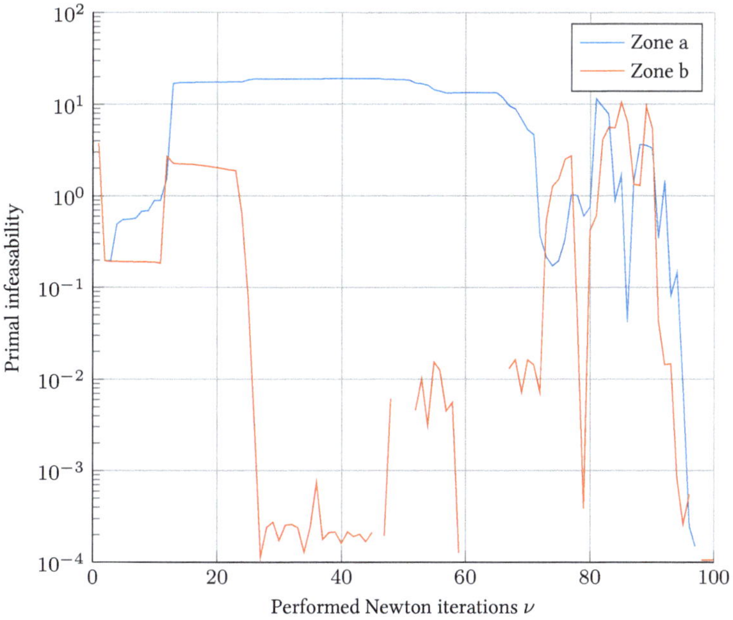

Figure 4.24: Primal infeasibility for current iteration ν. Note, that the infeasibility is not plotted if it drops below 10^{-4}.

Eight Node Network

The Eight Node Network did not show the expected convergence properties, see [MIJSH22]. In contrast to the Four Node Network, where convergence of the distributed solution was seen for a majority of the tested parameter sets, in the regarded Eight Node Network only a minority of the regarded parameter sets showed a convergence of the OCD approach toward the central solution. A potential explanation for this result is that in the Eight Node Network case, which in contrast to the Four Node Network includes the thermal pipeline models, Condition 3.2 might be regularly violated.

4.6 Summary and Discussion

This section summarizes and discusses the main findings of the performed simulations, demonstrating the applicability and limitations of the proposed TCS for CEPDHNs.

The application of the HPA in Case Study I showed how the uniform marginal prices obtained from the dispatch solution are calculated for a CEPDHN. Case Study II and III show how the ISOEMS fully utilizes the different flexibilities of the CEPDHN in a predictive manner, to maximize the welfare and thereby minimizing curtailment of RESs, and the losses occurring in the CEPDHN. The used flexibility originates from storage systems, thermal inertia of the DHN pipeline, as well as from demand side management of consumers in the EPN and the DHNs. Also, comparisons of the coordinated operation of the EPN with the DHN with a separated form of dispatch show the superiority of the coupled operation. Further, different price signals have been provided obtained from the TCS operation. The results of Case Study IV highlight that the proposed transactive control approach for CEPDHNs can also be used for further applications as operation cost minimization of a network operator by adapting the objective of the ISOEMS and the information exchange between the engaged entities. Further, these simulations show, that for large CEPDHNs, the approach, based on the detailed CEPDHN model presented in Chapter 2 leads to a high computational burden. This limits the real time capability of a central ISOEMS on a standard work station with of the shelf solvers. Case Study V demonstrated the application of the distributed rolling horizon approach based on OCD. The results showed expected convergence properties for a majority of the tested parameter sets for a small example DHN but the distributed solution does not converge toward the central results for the second tested Eight Node Network. In summary, these results presented different aspects of the CEPDHN operation performed by the TCS.

The application of the HPA in Case Study I showed the effectiveness of this pricing approach, in case that no market participant is able to exploit his market power. However in the case of large market power of single market participants further measures as regulation or grid reinforcements would be necessary to limit profits of respective actors.

The comparisons performed in Case Study II between the coordinated operation of the EPN and the DHN with a separated form of dispatch shows the superiority of the coupled operation, which is in line with the state of the art [ZWW$^+$21, WYJ$^+$16, LSF14]. Similar results provided in the literature showing the benefits of joint operation of CEPDHNs with flexible demand enhancing social welfare up to 16.6 % are provided in [SDWS17]. Also simulational results in [LWW$^+$19] show that an optimal joint operation of CEPDHNs leads to a 17.19 % RESs curtailment reduction. Note, that the superiority of joint CEPDHN operation not only stems from higher overall welfare but also from the absence of insecure network operation conditions[8]. This can be seen from the simulations provided in Case Study II, where a naive sequential dispatch of the EPN and the DHN led to failure of convergence of the optimization.

The simulational results provided in Case Study III show how important it is to take VMFDs into account when dispatching DHNs. Precisely, enabling VMFDs in the regarded scenario raises the accumulated social welfare by 20.72 %, as this allows to supply consumers with the lowest priced offers. To the best of the authors knowledge this aspect has so far not been sufficiently researched in the context of market-based DHN operation.

The findings from Case Study IV, demonstrating the high computational burden of the high detail DHN model, surpassing the 15 min time step intervalls for small prediction horizons below 6 time steps, align with statements on computational effort from [ZWW+21]. In this context, the interest in possibilities to speed up computations by parallelization techniques is straightforward. Regarding the results of the proposed OCD approach for DHNs, the increase in iterations ν needed to obtain the distributed solution compared to the central solution for the Four Node Network is in line with findings from [CNP02, NPC03]. These sources show, that the amount of iterations needed by the OCD approach are typically larger, as the search direction is calculated in an approximated form. Still, the single iterations will take less computation time for large scale problems, which is the basis for the calculational time reduction obtained from OCD. However, the limited convergence properties of the larger DHN solved by OCD in Case Study V showed, that the approach cannot be applied to general DHNs, in its current form.

Based on the aforementioned, the scalability of the overall TCS approach is clearly limited to DHNs with medium size, within the CEPDHNs , as the DHN optimization model comes along with a high computational burden. Based on the calculation times obtained from Case Study III and IV the possible maximum DHN size lies between 39 and 300 nodes on a standard work station with of the shelf solvers. On the other hand, the approach is expected to scale over large EPNs as calculation time reduction by distributed parallel optimization of EPNs based on OCD has been reported by multiple sources [NPC03, GHT16]. Thereby, it is important that the coupling condition has to be fulfilled for the respective decompositions.

All in all, the presented simulation results show the applicability of the presented TCS, to operate CEPDHNs in a technical secure[8] and efficient as well as welfare maximizing form.

5 Conclusion

The need for a carbon dioxide free and at the same time affordable energy supply creates a demand for new forms of network operation in EPNs and DHNs. As motivated in Chapter 1 desired properties of future CEPDHN operation approaches are the ability to coordinate a vast number of flexible network participants while achieving technical efficiency, economic optimality and secure[8] network operation. For that aim, existing economic and technical inefficiencies are to be reduced by designing market and control mechanisms in a joint form and by operating the physically coupled EPNs and DHNs in a coordinated form. Therefore, the objective and central contribution of this thesis are the design of a transactive control system for CEPDHNs. The TCS was designed with two hierarchy levels, comprising the EMSSAs on the first level and the ISOEMSs on the second level. Every EMSSA, representing an FNP, sends its bids or offers to the ISOEMS of its proper operational zone. As no further information besides the bids or offers is shared by the EMSSAs the approach enables a high level of privacy [KW16], while enabling end user participation [LLJZ20]. The ISOEMSs also obtain further power demand and supply forecasts and operational values as e.g. pump operation set points of their operational zone. Using all the provided information together with the contributed detailed CEPDHN model and the given operational limits, the ISOEMSs determine the optimal control values for the CEPDHNs. Thereby, the ISOEMSs achieve economic optimality by maximizing the social welfare, technical efficiency through the detailed CEPDHN model and secure[8] network operation by taking into account the operational limits. The overall approach represents a distributed rolling horizon procedure.

In contrast to the state of the art, the detailed network model considers all relevant physical effects of CEPDHN operation. These are variable mass flows with VMFDs, an exact representation of the pipeline friction factors, storage effects of pipelines, DPRs, pumps, and their respective control paths, thermal energy storage systems and lossy meshed AC EPNs. The comparatively high modeling detail of the presented CEPDHN model brings along two important features. First, it enables a technically efficient form of CEPDHN operation with low losses. Second, this modeling is the basis for enabling welfare maximization. In summary, the detailed CEPDHN model forms Contribution 2 of this work, bridging the Research Gap 1 a) b) stated in Section 1.2.5, see Figure 5.1.

As demonstrated in the results in Chapter 4, the detailed CEPDHN model brings along a high computational burden. Therefore, counteractive measures are taken in order to reduce the necessary computation times. First, the thermal pipeline models within the CEPDHN model are directly embedded into the rolling horizon approach. This is achieved by using precalculated mass flow parameters in the thermal pipeline model which avoid the need for integer variables, which would necessitate to solve a badly scaling MINLP problem at runtime. The presented approach precalculates the mass flow parameters after every optimization following the rolling horizon implementation, as explained in Section 3.3.1. Second, a distributed

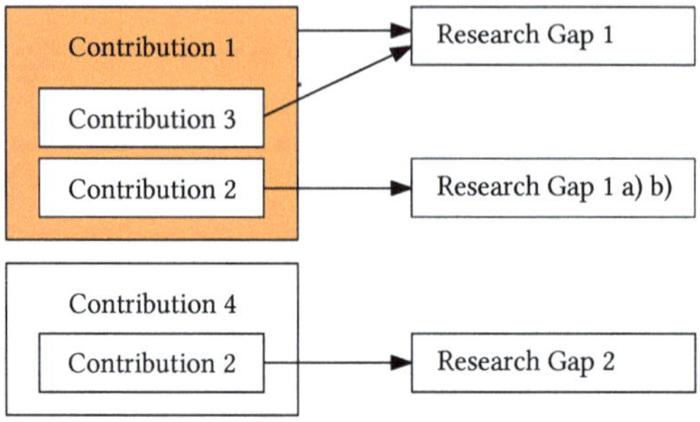

Figure 5.1: Relation between research gaps and contributions of this work.

rolling horizon approach is utilized which enables to parallelize the computations by multiple ISOEMSs. Therefore, different distributed optimization approaches have been compared. The optimality condition decomposition approach was selected due to the low amount of necessary information exchange between the ISOEMSs, the absence of tuning parameters, and its successful application within distributed optimization of EPNs and CEPDHNs. Building upon the aforementioned, it is expected, that large scale CEPDHNs and their respective network graphs can be decomposed into multiple subproblems by decomposing the CEPDHNs into multiple operational zones among electric tie lines. Every operational zone is controlled by its proper ISOEMS. When decomposing the initial central optimization problem into the small scale subproblems it is important to fulfill the conditions of the OCD approach, described in Section 3.3.2. At the same time, the coupling condition can be used as a quantitative criterion to compare different decompositions among each other. Visually speaking, stronger coupled subproblems, e.g. operational zones connected by more tie lines, will converge slower than weakly coupled subproblems. In certain cases, when the OCD conditions are not fulfilled, a distinct decomposition can lead to divergence or deviations to the central solution in general. So far the OCD approach has not been applied to DHNs, and distributed dispatch of DHNs in general represents Research Gap 2. Due to the strong coupling of the subproblems, originating from the decomposition of optimal dispatch problems of DHNs a specially fitted reformulation of the DHN model was necessary. This enabled to apply the OCD approach to a small scale DHN, which represents Contribution 4 of this thesis. However, the expected scalability of this approach is not provided for DHNs, as shown in Chapter 4.

Within Section 3.2.3 different pricing approaches have been outlaid and compared. The most efficient form of pricing, the locational marginal pricing approach can not be simply applied to DHNs as two LMPs are obtained for every FNP. Also, this work is contextualized within the German and thereby European energy market, where uniform marginal pricing is mostly applied[124]. Therefore, the novel hybrid pricing approach for CEPDHNs has been developed, which enables uniform marginal pricing in the case of uncongested network operation. Thereby, incentive compatibility is achieved in this scenario. When operational limits are reached, cases may occur, were some bids and offers are paid-as-bid and pay-as-bid vice versa. The necessary sum needed to pay these bids and offers is collected through a grid fee. As the approach inherently minimizes this sum, by maximizing the welfare the expected value

[124] Within every price zone.

of this grid fee is low. At the same time theoretically unbounded grid fees arising via uplift costs from redispatch are omitted, as operational limits of the network are considered during market clearing. Thereby, the hybrid pricing mechanism uses a price zone for the EPN and every DHN in order to account for different price levels of electric energy and heat. The presented novel pricing mechanism constitutes Contribution 3 of this work and represents an important part of in the transactive control system design, which yields Contribution 1. The interplay of, Contribution 1 and Contribution 3 help to fill Research Gap 1, which describes the missing coordination of market and control mechanisms within CEPDHN operation.

Figure 5.1 summarizes how the contributions of this work help to bridge the depicted research gap. The later comprises the insufficient coordination of market and control mechanisms for CEPDHN operation, the absence of detailed CEPDHN models needed for technical efficiency and welfare maximization, as well as the lack of approaches for distributed optimization of DHNs. The designed TCS, based on the novel CEPDHN model, enables secure[8], technically efficient and welfare maximizing operation of CEPDHNs. The simulational results presented in Chapter 4 illustrate benefits as loss reduction and flexibility provision. The later is achieved through storage capacities of DHN pipelines, thermal energy storage systems, battery storages, as well as demand side management in the DHN and the EPN. The highly coordinated operation of the EPN and the DHN leads to very low curtailment of RESs. Further, the case studys indicate the following three aspects: First, in line with the large amount of literature in this field, the results show, that coupled operation of the EPN and DHNs leads to the overall achievable welfare maximum. Second, a naive sequential market clearing of the electricity and the heat markets will lead to severe challenges in the network of the second cleared market. In the regarded scenarios no secure network operation condition could be provided in this case. Third, technical efficient and welfare maximizing operation of CEPDHNs is not possible with prevalent model simplifications found in current literature on optimal operation of CEPDHNs. Performed simulations showed that providing the possibility of mass flow direction changes in the network model led to a 20.72 % raise of the accumulated welfare. Also, variable mass flows are the basis for demand-side management and lowering heat losses in DHNs.

Several interesting aspects may be explored further by future work. These comprise the extension of the approach to further energy domains as for example gas networks. As gases are much more compressible than water, and this compressibility is actively used by gas network operators to store more or less energy within their networks this adds a further dimension to the regarded approach. Also the heating value of the transported gas changes in location and time, especially when different gases as natural gas, green gas and hydrogen are mixed.

Besides the approach could potentially be modified in order to enable a locational marginal pricing approach. However, this represents a nontrivial challenge as applying this pricing mechanism to DHNs brings along multiple difficulties as described in Section 3.2.3.

The development and implementation of transactive control approaches for multi-energy systems can reduce the costs of a future sustainable energy supply. This arises from the fact that these approaches can help to prevent grid reinforcements, lower operational costs through technical efficiency or the absence of uplift costs, and enable to achieve welfare maximizing operation.

A Determination of the Nodal Admittance Matrix

This section describes how the nodal admittance matrix $\underline{\boldsymbol{Y}}^{\mathrm{bus}}$ is calculated based on the π-equivalent impedance $\underline{Z}_e^{\pi}$ and admittance $\underline{Y}_e^{\pi}$, shown in Figure 2.3, for all transmission lines e, as well as the shunt admittance vector of all buses $\underline{\boldsymbol{Y}}^{\mathrm{sh}}$.

Based on the aforementioned, the transmission line admittance matrix $\underline{Y}_e^{\mathrm{trl}}$ of network edge e can be calculated as [ZMS11][125]:

$$\underline{Y}_e^{\mathrm{trl}} = \begin{bmatrix} \left.\frac{\underline{I}_e^{\mathrm{f}}}{\underline{V}_e^{\mathrm{f}}}\right|_{\underline{V}_e^{\mathrm{t}}=0} & \left.\frac{\underline{I}_e^{\mathrm{f}}}{\underline{V}_e^{\mathrm{t}}}\right|_{\underline{V}_e^{\mathrm{f}}=0} \\ \left.\frac{\underline{I}_e^{\mathrm{t}}}{\underline{V}_e^{\mathrm{f}}}\right|_{\underline{V}_e^{\mathrm{t}}=0} & \left.\frac{\underline{I}_e^{\mathrm{t}}}{\underline{V}_e^{\mathrm{t}}}\right|_{\underline{V}_e^{\mathrm{f}}=0} \end{bmatrix} = \begin{bmatrix} \underline{Y}_e^{\mathrm{ff}} & \underline{Y}_e^{\mathrm{ft}} \\ \underline{Y}_e^{\mathrm{tf}} & \underline{Y}_e^{\mathrm{tt}} \end{bmatrix} \tag{A.1}$$

$$= \begin{bmatrix} \frac{1}{\underline{Z}_e^{\pi}} + \frac{\underline{Y}_e^{\pi}}{2} & -\frac{1}{\underline{Z}_e^{\pi}} \\ -\frac{1}{\underline{Z}_e^{\pi}} & \frac{1}{\underline{Z}_e^{\pi}} + \frac{\underline{Y}_e^{\pi}}{2} \end{bmatrix}, \quad \forall\, e \in \mathbb{S}_e^{\mathrm{epn}} \tag{A.2}$$

where the superscripts ff, ft,tf, and ff of the admittances $\underline{Y}$ represent a sequence of the f and t superscripts identifying the respective currents and voltages at the *from* or *to* end of the transmission line, see Figure 2.3. The auxiliary admittance column vectors $\underline{\boldsymbol{Y}}^{\mathrm{ff}}$, $\underline{\boldsymbol{Y}}^{\mathrm{ft}}$, $\underline{\boldsymbol{Y}}^{\mathrm{tf}}$, and $\underline{\boldsymbol{Y}}^{\mathrm{tt}}$ are set up as:

$$\underline{\boldsymbol{Y}}^{\mathrm{ff}} = \underline{\boldsymbol{Y}}^{\mathrm{tt}} = \left[\underline{Y}_1^{\mathrm{ff}}, \dots, \underline{Y}_e^{\mathrm{ff}}, \dots, \underline{Y}_{\mathrm{n^{ce}}}^{\mathrm{ff}}\right] \tag{A.3}$$

$$\underline{\boldsymbol{Y}}^{\mathrm{ft}} = \underline{\boldsymbol{Y}}^{\mathrm{tf}} = \left[\underline{Y}_1^{\mathrm{ft}}, \dots, \underline{Y}_e^{\mathrm{ft}}, \dots, \underline{Y}_{\mathrm{n^{ce}}}^{\mathrm{ft}}\right] \tag{A.4}$$

with the cardinality of the EPN edges $\mathrm{n^{ce}} = \left|\mathbb{S}_e^{\mathrm{epn}}\right|$.

The nodal admittance matrix $\underline{\boldsymbol{Y}}^{\mathrm{bus}}$ of the entire EPN is then calculated by [ZMS11]:

$$\underline{\boldsymbol{Y}}^{\mathrm{f}} = \mathrm{diag}\left(\underline{\boldsymbol{Y}}^{\mathrm{ff}}\right)\left(\underline{\boldsymbol{A}}^{\mathrm{epn},-}\right)^{\top} + \mathrm{diag}\left(\underline{\boldsymbol{Y}}^{\mathrm{ft}}\right)\left(\underline{\boldsymbol{A}}^{\mathrm{epn},+}\right)^{\top} \tag{A.5}$$

$$\underline{\boldsymbol{Y}}^{\mathrm{t}} = \mathrm{diag}\left(\underline{\boldsymbol{Y}}^{\mathrm{tf}}\right)\left(\underline{\boldsymbol{A}}^{\mathrm{epn},-}\right)^{\top} + \mathrm{diag}\left(\underline{\boldsymbol{Y}}^{\mathrm{tt}}\right)\left(\underline{\boldsymbol{A}}^{\mathrm{epn},+}\right)^{\top} \tag{A.6}$$

$$\underline{\boldsymbol{Y}}^{\mathrm{bus}} = \underline{\boldsymbol{A}}^{\mathrm{epn},-}\underline{\boldsymbol{Y}}^{\mathrm{f}} + \underline{\boldsymbol{A}}^{\mathrm{epn},+}\underline{\boldsymbol{Y}}^{\mathrm{t}}\mathrm{diag}\left(\underline{\boldsymbol{Y}}^{\mathrm{sh}}\right) \tag{A.7}$$

where the matrices $\underline{\boldsymbol{A}}^{\mathrm{epn},+}$ and $\underline{\boldsymbol{A}}^{\mathrm{epn},-}$ are given in (2.14).

[125] The negative sign of the admittances on the secondary diagonals originates from the opposed directions of flow and potential difference, see Figure 2.3.

B Possible Forms of Decompositions

The aim of the following explanations is to show that insufficient degrees of freedom for at least one zone can arise, when applying different forms of decomposition. Different possible forms of decomposing a DHN into similarly sized subproblems as provided initially in Figure 3.10 in the Example 3.3 are provided in the following. Four different zone partitioning possibilities are shown in Figure B.1 by the different Borders 1 to 4 used to decompose the small DHN into the two subproblems, Zone a and Zone b. Note, that for every border shown in Figure B.1 the complicating constraint arising from the border loop, exemplary stated in (B.2), can be either assigned to Zone a or Zone b. Thus eight possible forms to decompose the original problem can be distinguished in total in this case. Since an example based on the decomposition by Border 2 has already been provided in Example 3.3, which is similar to the decomposition by Border 1, we will first demonstrate the effects of a decomposition by Border 4, which is again analog to the procedure using Border 3. The complicating constraints arising from applying the decomposition defined by Border 4 in Zone b are given by:

$$\overline{\dot{m}_1} - \dot{m}_2 + \dot{m}_6 = 0 \tag{B.1}$$

$$\overline{\Delta p_1} + \Delta p_2 + \Delta p_3 - \overline{\Delta p_4} = 0 \tag{B.2}$$

Thereby, the complication loop equation (B.2) was assigned to Zone b. The second hydraulic node equation is:

$$\dot{m}_2 - \dot{m}_3 - \dot{m}_6 = 0 \tag{B.3}$$

Note, that from (B.1) and (B.3) it is apparent, that $\overline{\dot{m}_1} = \dot{m}_3$. Based on this from (3.46d) for e_3 it can be seen, that also Δp_3 is defined in this case. Using this information, from (B.2) it can be seen, that also the value of Δp_2 is determined, and with (3.46d) for e_2 this also incorporates $\dot{m}_2$. Since thereby, $\dot{m}_6$ is also known from (B.1), the system of equations once more has zero degrees of freedom[126].

Note, that if the network operation would limit only one further degree of freedom, e.g. through a set operation point of a pump, e.g. defining Δp_2, then the both subproblems would have zero degrees of freedom in this case.

However, for completeness it should also be noted, that using a decomposition based on Border 2, assigning the complicating constraint of the border loop equation to Zone a would not fully predefine the flows and pressures in either of the zones. However, in contrast to OCD applied to EPNs, see Example 3.4, Zone b would have no possibility to change the flows to Zone a, as they are predefined in the complicating constraints originating from the hydraulic node equations (3.46b) of Node 2 and Node 4.

[126] Note, that the same occurs with the systems of equations of Zone a if the complicating constraint of the border loop equation, here (B.2), would have been assigned to Zone a.

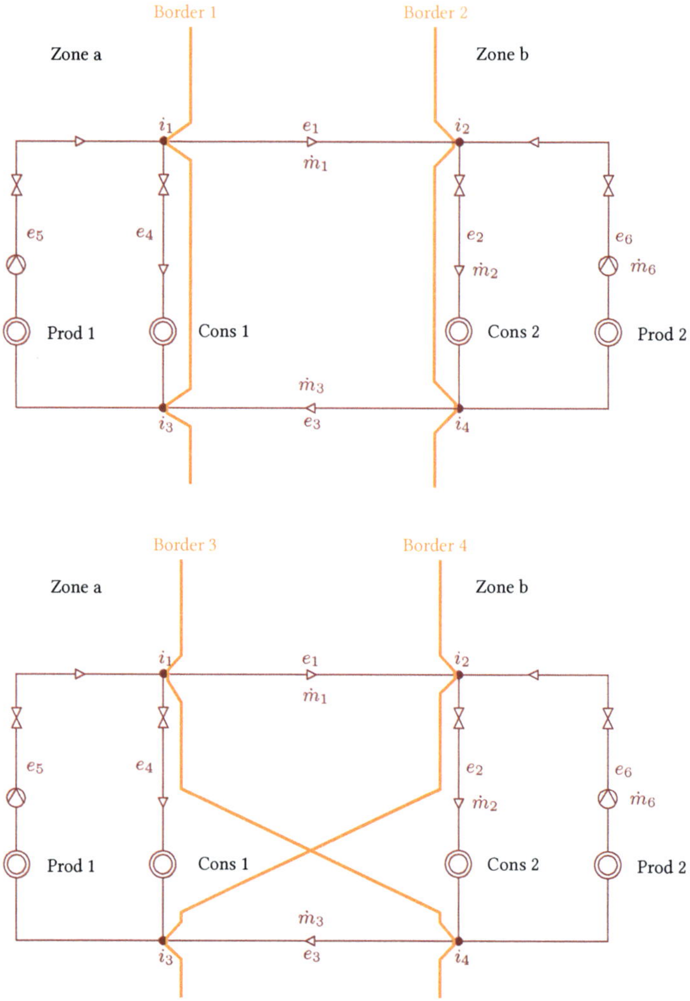

Figure B.1: Example DHN decomposed into Zone a and Zone b by four possible forms (Border 1 to Border 4) to obtain similarly sized subproblems.

In summary, the provided forms of decomposition showed that in most cases using OCD to the example DHN showed in Figure B.1, the system of equations of one zone had zero degrees of freedom. If only a single degree of freedom of the system of hydraulic model equations, is reduced by the form of network operation, both zones can have zero degrees of freedom within their system of equations. This shows, that applying OCD to DHNs can be challenging. These aspects are not relevant when applying OCD to EPNs, see Example 3.4.

C The Border Loop Equation

This appendix provides further insight on the pressure drop loop equation, which is replaced by the model reformulation. The pressure drop loop equation (2.24) is replaced by the following equations (C.1) for border loops in the approach presented in Section 3.3.2:

$$\left(\sum_{e \in \mathbb{S}_e^{\text{dhn,za}}, e \notin \mathbb{S}_e^{\text{dhn,be,za}}} \sum_{i \in \mathbb{S}_i^{\text{dhn,bi,za}}} |B_{l,e}||A_{i,e}^{\text{dhn}}| p_{i,k} \right) = \Delta p^{\text{pre}}, \qquad \forall\, l \in \mathbb{S}_l^{\text{dhn,bl,za}}, k \in \mathbb{S}_k \tag{C.1}$$

with the set ob all border loops in Zone a stated in $\mathbb{S}_l^{\text{dhn,bl,za}}$. In order to guarantee, that the pressure difference aligns with the physical flow direction, an additional inequality (C.2) is used, as stated below:

$$\begin{aligned} &\left(\sum_{e \in \mathbb{S}_e^{\text{dhn,za}}, e \notin \mathbb{S}_e^{\text{dhn,be,za}}} \sum_{i \in \mathbb{S}_i^{\text{dhn,sn,bi,za}}} |B_{l,e}||A_{i,e}^{\text{dhn}}| p_{i,k} \right) \\ > &\left(\sum_{e \in \mathbb{S}_e^{\text{dhn,za}}, e \notin \mathbb{S}_e^{\text{dhn,be,za}}} \sum_{i \in \mathbb{S}_i^{\text{dhn,rn,bi,za}}} |B_{l,e}||A_{i,e}^{\text{dhn}}| p_{i,k} \right), \ \forall\, l \in \mathbb{S}_l^{\text{dhn,bl,za}}, k \in \mathbb{S}_k \end{aligned} \tag{C.2}$$

Remark C.1:

This formulation is equivalent to considering the differential pressure loop equation (2.24) *of the central optimization problem, in both zones of the decomposed OCD form, for each border loop. This can be formulated as follows:*

$$\sum_{e\in\mathbb{S}_e^{\mathrm{dhn}}} B_{l,e}\Delta p_{e,k} = 0 \tag{C.3a}$$

$$= \sum_{e\in\mathbb{S}_e^{\mathrm{dhn,za}},e\notin\mathbb{S}_e^{\mathrm{dhn,be,za}}} B_{l,e}\Delta p_{e,k} + \sum_{e\in\mathbb{S}_e^{\mathrm{dhn,zb}},e\notin\mathbb{S}_e^{\mathrm{dhn,be,zb}}} B_{l,e}\Delta p_{e,k} + \sum_{e\in\mathbb{S}_e^{\mathrm{dhn,be,za}},e\in\mathbb{S}_e^{\mathrm{dhn,be,zb}}} B_{l,e}\Delta p_{e,k} \tag{C.3b}$$

$$= \sum_{e\in\mathbb{S}_e^{\mathrm{dhn,za}},e\notin\mathbb{S}_e^{\mathrm{dhn,be,za}}} B_{l,e}\Delta p_{e,k} + \sum_{e\in\mathbb{S}_e^{\mathrm{dhn,zb}},e\notin\mathbb{S}_e^{\mathrm{dhn,be,zb}}} B_{l,e}\Delta p_{e,k} + 2\sum_{e\in\mathbb{S}_e^{\mathrm{dhn,sn,be,za}},e\in\mathbb{S}_e^{\mathrm{dhn,sn,be,zb}}} B_{l,e}\Delta p_{e,k} \tag{C.3c}$$

$$= \left(\sum_{e\in\mathbb{S}_e^{\mathrm{dhn,za}},e\notin\mathbb{S}_e^{\mathrm{dhn,be,za}}} \sum_{i\in\mathbb{S}_i^{\mathrm{dhn,bi,za}}} |B_{l,e}||A_{i,e}^{\mathrm{dhn}}|p_{i,k}\right) + \left(\sum_{e\in\mathbb{S}_e^{\mathrm{dhn,zb}},e\notin\mathbb{S}_e^{\mathrm{dhn,be,zb}}} \sum_{i\in\mathbb{S}_i^{\mathrm{dhn,bi,zb}}} |B_{l,e}||A_{i,e}^{\mathrm{dhn}}|p_{i,k}\right) \tag{C.3d}$$

$$\forall\, l \in \mathbb{S}_l^{\mathrm{dhn,bl,za}}, k \in \mathbb{S}_k$$

The above is equal to equation (C.1) *as the constant differential pressure* Δp^{pre} *in equation* (C.1) *is defined as follows for a given border loop* l *of zone a:*

$$\Delta p^{\mathrm{pre}} = \left(\sum_{e\in\mathbb{S}_e^{\mathrm{dhn,zb}},e\notin\mathbb{S}_e^{\mathrm{dhn,be,zb}}} \sum_{i\in\mathbb{S}_i^{\mathrm{dhn,bi,zb}}} |B_{l,e}||A_{i,e}^{\mathrm{dhn}}|p_{i,k}\right), \quad l \in \mathbb{S}_l^{\mathrm{dhn,bl,za}}, k \in \mathbb{S}_k \tag{C.4}$$

Strictly regarded, the replacement of the pressure drop border loop equation (2.24) by the above does not yet resemble a decomposition of the original central problem into decomposed smaller problems, as usually performed by OCD, as no complicating constraints are introduced here. Much more, this approach enables the pressure potentials to remain variable while using the OCD method, as stated in Section 3.3.2.

References

Public References

[AB19] ALRUMAYH, O.; BHATTACHARYA, K.: Flexibility of Residential Loads for Demand Response Provisions in Smart Grid. In: *IEEE Transactions on Smart Grid* 10 (2019), Nr. 6, S. 6284–6297

[ACP+14] AUSUBEL, L. M.; CRAMTON, P.; PYCIA, M.; ROSTEK, M.; WERETKA, M.: Demand Reduction and Inefficiency in Multi-Unit Auctions. In: *The Review of Economic Studies* 81 (2014), Nr. 4, S. 1366–1400

[Ago19] AGORA: Verteilnetzausbau für die Energiewende - Elektromobilität im Fokus / Agora Verkehrswende, Agora Energiewende, Regulatory Assistance Project. 2019. – Forschungsbericht

[ALFV19] ABRISHAMBAF, O.; LEZAMA, F.; FARIA, P.; VALE, Z.: Towards transactive energy systems: An analysis on current trends. In: *Energy Strategy Reviews* 26 (2019), S. 100418

[Arn11] ARNOLD, M. J.: *On predictive control for coordination in multi-carrier energy systems*, ETH Zurich, Diss., 2011

[ARRB16] AMROUCHE, S. O.; REKIOUA, D.; REKIOUA, T.; BACHA, S.: Overview of energy storage in renewable energy systems. In: *International journal of hydrogen energy* 41 (2016), Nr. 45, S. 20914–20927

[AW14b] ADIKA, C. O.; WANG, L.: Demand-side bidding strategy for residential energy management in a smart grid environment. In: *IEEE Transactions on Smart Grid* 5 (2014), Nr. 4, S. 1724–1733

[BBM17] BORRELLI, F.; BEMPORAD, A.; MORARI, M.: *Predictive control for linear and hybrid systems.* Cambridge University Press, 2017

[BCDC18] BEHBOODI, S.; CHASSIN, D. P.; DJILALI, N.; CRAWFORD, C.: Transactive control of fast-acting demand response based on thermostatic loads in real-time retail electricity markets. In: *Applied Energy* 210 (2018), S. 1310–1320

[Ben91] BENONYSSON, A.: *Dynamic modelling and operational optimization of district heating systems*, Technical University of Denmark, Lyngby, Diss., 1991

[BHH+12] BOZCHALUI, M. C.; HASHMI, S. A.; HASSEN, H.; CANIZARES, C. A.; BHATTACHARYA, K.: Optimal operation of residential energy hubs in smart grids. In: *IEEE Transactions on Smart Grid* 3 (2012), Nr. 4, S. 1755–1766

[BHK+02] Bøhm, B.; Ha, S.-k.; Kim, W.-t.; Kim, B.-k.; Koljonen, T.; Larsen, H. V.; Lucht, M.; Park, Y.-s.; Sipilä, K.; Wigbels, M. u. a.: Simple models for operational optimisation / IEA. 2002. – Forschungsbericht. – 0010 S.

[BPC+11] Boyd, S.; Parikh, N.; Chu, E.; Peleato, B.; Eckstein, J.: Distributed optimization and statistical learning via the alternating direction method of multipliers. In: *Foundations and Trends® in Machine Learning* 3 (2011), Nr. 1, S. 1–122

[Brk11] Brkić, D.: Review of explicit approximations to the Colebrook relation for flow friction. In: *Journal of Petroleum Science and Engineering* 77 (2011), Nr. 1, S. 34–48

[Bun18b] Bundesnetzagentur, B.: Monitoring report 2018 / Bundesnetzagentur. Author Bonn, 2018. – Forschungsbericht

[Bun19a] Bundesnetzagentur | SMARD.de: *Marktdaten.* 2019. – [Online; accessed: 05.11.2019 and 30.10.2021]

[Bun19b] Bundesrechnungshof: *Bericht an das Bundesministerium für Wirtschaft und Energie nach § 88 Abs. 2 BHO zur Prüfung von Maßnahmen zum Netzausbau für die Energiewende.* May 2019

[Bun22] Bundesnetzagentur: Monitoringbericht 2021 / Bundesnetzagentur, Bundeskartellamt. 2022. – Forschungsbericht

[CBS82] Caramanis, M. C.; Bohn, R. E.; Schweppe, F. C.: Optimal spot pricing: Practice and theory. In: *IEEE Transactions on Power Apparatus and Systems* (1982), Nr. 9, S. 3234–3245

[CCMGB06] Conejo, A. J.; Castillo, E.; Minguez, R.; Garcia-Bertrand, R.: *Decomposition techniques in mathematical programming: engineering and science applications.* Springer Science & Business Media, 2006

[CKC+14] Chakrabarti, S.; Kraning, M.; Chu, E.; Baldick, R.; Boyd, S.: Security constrained optimal power flow via proximal message passing. In: *2014 Clemson University Power Systems Conference* IEEE, 2014, S. 1–8

[CMT+17] Christidis, A.; Mollenhauer, E.; Tsatsaronis, G.; Schuchardt, G. K.; Holler, S.; Böttger, D.; Bruckner, T.: EnEff-Wärme: Einsatz von Wärmespeichern und Power-to-Heat-Anlagen in der Fernwärmeerzeugung / Technische Universität Berlin. 2017. – Forschungsbericht

[CN17] Christian Nabe, M. D. U. H. N. G. Marie-Louise Arlt A. Marie-Louise Arlt: Smart-Market-Design in deutschen Verteilnetzen / Agora. 03/2017. – Forschungsbericht

[CNP02] Conejo, A. J.; Nogales, F. J.; Prieto, F. J.: A decomposition procedure based on approximate Newton directions. In: *Mathematical programming* 93 (2002), Nr. 3, S. 495–515

[Cor20] Corporation, G. D.: *GAMS documentation: Distribution 28.1.0.* July 2020

[Cou15] Council, T. G. A.: GridWise Transactive Energy Framework / The GridWise Architecture Council. 2015. – Forschungsbericht

[CWL+18] Chen, Y.; Wei, W.; Liu, F.; Sauma, E. E.; Mei, S.: Energy trading and market equilibrium in integrated heat-power distribution systems. In: *IEEE Transactions on Smart Grid* 10 (2018), Nr. 4, S. 4080–4094

[CWW+18a] Cao, Y.; Wei, W.; Wang, J.; Mei, S.; Shafie-khah, M.; Catalao, J. P.: Capacity planning of energy hub in multi-carrier energy networks: A data-driven robust stochastic programming approach. In: *IEEE Transactions on Sustainable Energy* 11 (2018), Nr. 1, S. 3–14

[CWW+18b] Cao, Y.; Wei, W.; Wu, L.; Mei, S.; Shahidehpour, M.; Li, Z.: Decentralized operation of interdependent power distribution network and district heating network: A market-driven approach. In: *IEEE Transactions on Smart Grid* 10 (2018), Nr. 5, S. 5374–5385

[ÇYÇ04] Çomakli, K.; Yüksel, B.; Çomakli, Ö.: Evaluation of energy and exergy losses in district heating network. In: *Applied thermal engineering* 24 (2004), Nr. 7, S. 1009–1017

[CYW20] Cai, H.; You, S.; Wu, J.: Agent-based distributed demand response in district heating systems. In: *Applied Energy* 262 (2020), S. 114403

[DAM19] Dénarié, A.; Aprile, M.; Motta, M.: Heat transmission over long pipes: New model for fast and accurate district heating simulations. In: *Energy* 166 (2019), S. 267–276

[DDKB+20] Davies, J.; Dolci, F.; Klassek-Bajorek, D.; Ortiz Cebolla, R.; Weidner, E.: Current status of chemical energy storage technologies. In: *Publications Office of the European Union: Luxembourg* (2020)

[Des69] Desoer, C. A.: *Basic circuit theory.* McGraw-Hill, 1969

[DF05] Ding, F.; Fuller, J. D.: Nodal, uniform, or zonal pricing: distribution of economic surplus. In: *IEEE transactions on Power Systems* 20 (2005), Nr. 2, S. 875–882

[DF14] Daniel Fürstenwerth, C. R. P. H. Norman Gerhardt G. Norman Gerhardt: Power-to-Heat zur Integration von ansonsten abgeregeltem Strom aus Erneuerbaren Energien / Agora Energiewende. 2014. – Forschungsbericht

[DFH09] Diehl, M.; Ferreau, H. J.; Haverbeke, N.: Efficient Numerical Methods for Nonlinear MPC and Moving Horizon Estimation. In: *Nonlinear Model Predictive Control.* Springer Berlin Heidelberg, 2009

[DHN] *Grundfos, EInstufige Blockpumpen, DHN Pump NB 100-250/205 BAF2AESBAQEUW1.* Online. `https://product-selection.grundfos.com/de/products/nb-nbe-nbe-series-2000/nb/nb-100-250205-98777276fltab=variant-curves&pumpsystemid=1434891109`. – Accesed 17.11.2021, 11:50

[DLS+19] Deng, L.; Li, Z.; Sun, H.; Guo, Q.; Xu, Y.; Chen, R.; Wang, J.; Guo, Y.: Generalized locational marginal pricing in a heat-and-electricity-integrated market. In: *IEEE Transactions on Smart Grid* 10 (2019), Nr. 6, S. 6414–6425

[DTKM18] Deb, S.; Tammi, K.; Kalita, K.; Mahanta, P.: Impact of electric vehicle charging station load on distribution network. In: *Energies* 11 (2018), Nr. 1, S. 178

[EFS03] Engell, S.; Frehse, G.; Schnieder, E.: *Modelling, analysis and design of hybrid systems*. Bd. 279. Springer, 2003

[Ers14] Erseghe, T.: Distributed optimal power flow using ADMM. In: *IEEE transactions on power systems* 29 (2014), Nr. 5, S. 2370–2380

[Ers15] Erseghe, T.: A distributed approach to the OPF problem. In: *EURASIP Journal on Advances in Signal Processing* 2015 (2015), Nr. 1, S. 1–13

[Fyk19] Fyke, A.: The fall and rise of gravity storage technologies. In: *Joule* 3 (2019), Nr. 3, S. 625–630

[GA07] Geidl, M.; Andersson, G.: Optimal power flow of multiple energy carriers. In: *IEEE Transactions on Power Systems* 22 (2007), Nr. 1, S. 145–155

[GCHM19] Good, N.; Ceseña, E. A. M.; Heltorp, C.; Mancarella, P.: A transactive energy modelling and assessment framework for demand response business cases in smart distributed multi-energy systems. In: *Energy* 184 (2019), S. 165–179

[GCM17] Good, N.; Ceseña, E. A. M.; Mancarella, P.: Ten questions concerning smart districts. In: *Building and Environment* 118 (2017), S. 362–376

[GECC18] Gomez-Exposito, A.; Conejo, A. J.; Canizares, C.: *Electric energy systems: analysis and operation*. CRC press, 2018

[GHT15] Guo, J.; Hug, G.; Tonguz, O.: Impact of partitioning on the performance of decomposition methods for AC Optimal Power Flow. In: *Innovative Smart Grid Technologies Conference (ISGT), 2015 IEEE Power & Energy Society* IEEE, 2015, S. 1–5

[GHT16] Guo, J.; Hug, G.; Tonguz, O. K.: Intelligent partitioning in distributed optimization of electric power systems. In: *IEEE Transactions on Smart Grid* 7 (2016), Nr. 3, S. 1249–1258

[GHT17] Guo, J.; Hug, G.; Tonguz, O. K.: A case for nonconvex distributed optimization in large-scale power systems. In: *IEEE Transactions on Power Systems* 32 (2017), Nr. 5, S. 3842–3851

[GHVO96] Guo, T.; Henwood, M. I.; Van Ooijen, M.: An algorithm for combined heat and power economic dispatch. In: *IEEE Transactions on Power Systems* 11 (1996), Nr. 4, S. 1778–1784

[Glü85b] Glück, B.: *Heizwassernetze für Wohn- und Industriegebiete*. Verlag für Bauwesen, 1985

[Gmb18] GmbH, E. N.: *Lastprofile, Temperaturtabellen.* www.ednetze.de. Version: 2018. – [Online; accessed: 27.04.2018]

[Grü17] Grüne, L.; Pannek, J. (Hrsg.): *Nonlinear Model Predictive Control : Theory and Algorithms.* 2nd ed. 2017. Cham: Springer, 2017

[GS21a] Giunta, F.; Sawalha, S.: Techno-economic analysis of heat recovery from supermarket's CO2 refrigeration systems to district heating networks. In: *Applied Thermal Engineering* 193 (2021), S. 117000

[GT17b] Guney, M. S.; Tepe, Y.: Classification and assessment of energy storage systems. In: *Renewable and Sustainable Energy Reviews* 75 (2017), S. 1187–1197

[GV19] Guelpa, E.; Verda, V.: Thermal energy storage in district heating and cooling systems: A review. In: *Applied Energy* 252 (2019), S. 113474

[GWL+17] Gu, W.; Wang, J.; Lu, S.; Luo, Z.; Wu, C.: Optimal operation for integrated energy system considering thermal inertia of district heating network and buildings. In: *Applied energy* 199 (2017), S. 234–246

[GYYW21] Gan, W.; Yan, M.; Yao, W.; Wen, J.: Peer to peer transactive energy for multiple energy hub with the penetration of high-level renewable energy. In: *Applied Energy* 295 (2021), S. 117027

[HAH17] Heijde, B. van d.; Aertgeerts, A.; Helsen, L.: Modelling steady-state thermal behaviour of double thermal network pipes. In: *International Journal of Thermal Sciences* 117 (2017), S. 316–327

[Ham20] Hampel, S.: *Optimal Cooperation of Microgrids Connected on Transmission Level*, IRS, KIT, Diplomarbeit, 2020

[HCC+12] Heydt, G. T.; Chowdhury, B. H.; Crow, M. L.; Haughton, D.; Kiefer, B. D.; Meng, F.; Sathyanarayana, B. R.: Pricing and control in the next generation power distribution system. In: *IEEE Transactions on Smart Grid* 3 (2012), Nr. 2, S. 907–914

[HFT+17] Heijde, B. van d.; Fuchs, M.; Tugores, C. R.; Schweiger, G.; Sartor, K.; Basciotti, D.; Müller, D.; Nytsch-Geusen, C.; Wetter, M.; Helsen, L.: Dynamic equation-based thermo-hydraulic pipe model for district heating and cooling systems. In: *Energy Conversion and Management* 151 (2017), S. 158–169

[HGA09] Hug-Glanzmann, G.; Andersson, G.: Decentralized optimal power flow control for overlapping areas in power systems. In: *IEEE Transactions on Power Systems* 24 (2009), Nr. 1, S. 327–336

[HL12] Holmberg, P.; Lazarczyk, E.: Congestion management in electricity networks: Nodal, zonal and discriminatory pricing. (2012)

[HLW17] Huang, J.; Li, Z.; Wu, Q.: Coordinated dispatch of electric power and district heating networks: A decentralized solution using optimality condition decomposition. In: *Applied Energy* 206 (2017), S. 1508 – 1522

[HTWL19] Huang, S.; Tang, W.; Wu, Q.; Li, C.: Network constrained economic dispatch of integrated heat and electricity systems through mixed integer conic programming. In: *Energy* 179 (2019), S. 464–474

[Ick95] Icking, M.: *Zur Modellierung des dynamischen Betriebs von Fernwärmesystemen.* Hartung-Gorre, 1995

[KAA$^+$19] Kougias, I.; Aggidis, G.; Avellan, F.; Deniz, S.; Lundin, U.; Moro, A.; Muntean, S.; Novara, D.; Pérez-Díaz, J. I.; Quaranta, E. u. a.: Analysis of emerging technologies in the hydropower sector. In: *Renewable and Sustainable Energy Reviews* 113 (2019), S. 109257

[Kel13] Keles, D.: *Uncertainties in energy markets and their consideration in energy storage evaluation*, Karlsruhe Institute of Technology, Diss., 2013

[Kie16] Kiel, S.: *Küstenkraftwerk K.I.E.L., Kiels intelligente Energie-Lösung.* Dezember 2016

[KMHG21] Kaisermayer, V.; Muschick, D.; Horn, M.; Gölles, M.: Operation of coupled multi-owner district heating networks via distributed optimization. In: *Energy Reports* 7 (2021), S. 273–281

[Köc00] Köcher, R.: *Beitrag zur Berechnung und Auslegung von Fernwärmenetzen*, TU Berlin, Diss., 2000

[Kon18] Konstantin, P.: *Praxisbuch der Fernwärmeversorgung: Systeme, Netzaufbauvarianten, Kraft-Wärme-Kopplung, Kostenstrukturen und Preisbildung.* Springer-Verlag, 2018

[KP17] Kiesel, R.; Paraschiv, F.: Econometric analysis of 15-minute intraday electricity prices. In: *Energy Economics* 64 (2017), S. 77–90

[Kun93] Kundur, P.: *Power system stability and control.* [Nachdr.]. New York [u.a.]: McGraw-Hill, 1993

[KW16] Kok, K.; Widergren, S.: A society of devices: Integrating intelligent distributed resources with transactive energy. In: *IEEE Power and Energy Magazine* 14 (2016), Nr. 3, S. 34–45

[LFZC16] Li, J.; Fang, J.; Zeng, Q.; Chen, Z.: Optimal operation of the integrated electrical and heating systems to accommodate the intermittent renewable sources. In: *Applied Energy* 167 (2016), 244 - 254

[LGZ$^+$19] Lu, S.; Gu, W.; Zhou, S.; Yu, W.; Yao, S.; Pan, G.: High-resolution modeling and decentralized dispatch of heat and electricity integrated energy system. In: *IEEE Transactions on Sustainable Energy* 11 (2019), Nr. 3, S. 1451–1463

[Liu13] Liu, X.: *Combined analysis of electricity and heat networks*, Cardiff University, Diss., 2013

[LLJZ20] Li, S.; Lian, J.; J. Conejo, A.; Zhang, W.: Transactive Energy Systems: The Market-Based Coordination of Distributed Energy Resources. In: *IEEE Control Systems Magazine* 40 (2020), August, Nr. 4, S. 26–52

[LMMD10] Lund, H.; Möller, B.; Mathiesen, B.; Dyrelund, A.: The role of district heating in future renewable energy systems. In: *Energy* 35 (2010), Nr. 3, S. 1381 – 1390

[LSF14] Liu, M.; Shi, Y.; Fang, F.: Combined cooling, heating and power systems: A survey. In: *Renewable and Sustainable Energy Reviews* 35 (2014), S. 1–22

[LSG+17] Li, H.; Svendsen, S.; Gudmundsson, O.; Kuosa, M.; Rämä, M.; Sipilä, K.; Blesl, M.; Broydo, M.; Stehle, M.; Pesch, R. u. a.: Future low temperature district heating design guidebook: Final Report of IEA DHC Annex TS1. Low Temperature District Heating for Future Energy Systems / International Energy Agency. International Energy Agency, 2017. – Forschungsbericht

[LSZW15] Li, H.; Sun, Q.; Zhang, Q.; Wallin, F.: A review of the pricing mechanisms for district heating systems. In: *Renewable and Sustainable Energy Reviews* 42 (2015), S. 56–65

[Lun14a] Lunze, J.: *Control theory of digitally networked dynamic systems.* Springer, 2014

[Lun14b] Lunze, J.: *Regelungstechnik 1 : Systemtheoretische Grundlagen, Analyse und Entwurf einschleifiger Regelungen.* 10., aktualisierte Aufl. 2014. Berlin, Heidelberg: Springer Vieweg, 2014

[Lun14c] Lunze, J.: *Regelungstechnik 2 : Mehrgrößensysteme, Digitale Regelung.* 8., überarb. Aufl. 2014. Berlin, Heidelberg: Springer Vieweg, 2014

[LWHZ17] Liu, Z.; Wu, Q.; Huang, S.; Zhao, H.: Transactive energy: A review of state of the art and implementation. In: *IEEE PowerTech Manchester* IEEE, 2017, S. 1–6

[LWJB16] Liu, X.; Wu, J.; Jenkins, N.; Bagdanavicius, A.: Combined analysis of electricity and heat networks. In: *Applied Energy* 162 (2016), S. 1238–1250

[LWM+18] Li, R.; Wei, W.; Mei, S.; Hu, Q.; Wu, Q.: Participation of an energy hub in electricity and heat distribution markets: An MPEC approach. In: *IEEE Transactions on Smart Grid* 10 (2018), Nr. 4, S. 3641–3653

[LWS+15] Li, Z.; Wu, W.; Shahidehpour, M.; Wang, J.; Zhang, B.: Combined Heat and Power Dispatch Considering Pipeline Energy Storage of District Heating Network. In: *IEEE Transactions on Sustainable Energy* 7 (2015), Jan, Nr. 1, S. 12–22

[LWW+15] Li, Z.; Wu, W.; Wang, J.; Zhang, B.; Zheng, T.: Transmission-constrained unit commitment considering combined electricity and district heating networks. In: *IEEE Transactions on Sustainable Energy* 7 (2015), Nr. 2, S. 480–492

[LWW18] Liu, N.; Wang, J.; Wang, L.: Hybrid energy sharing for multiple microgrids in an integrated heat-electricity energy system. In: *IEEE Transactions on Sustainable Energy* 10 (2018), Nr. 3, S. 1139–1151

[LWW+19] Lin, C.; Wu, W.; Wang, B.; Shahidehpour, M.; Zhang, B.: Joint Commitment of Generation Units and Heat Exchange Stations for Combined Heat and Power Systems. In: *IEEE Transactions on Sustainable Energy* 11 (2019), July, Nr. 3, S. 1118 – 1127

[LWZS17] Lin, C.; Wu, W.; Zhang, B.; Sun, Y.: Decentralized solution for combined heat and power dispatch through benders decomposition. In: *IEEE Transactions on sustainable energy* 8 (2017), Nr. 4, S. 1361–1372

[MAFH21] Miyasawa, A.; Akira, S.; Fujimoto, Y.; Hayashi, Y.: Spatial demand forecasting based on smart meter data for improving local energy self-sufficiency in smart cities. In: *IET Smart Cities* 3 (2021), Nr. 2, S. 107–120

[Man14] Mancarella, P.: MES (multi-energy systems): An overview of concepts and evaluation models. In: *Energy* 65 (2014), S. 1–17

[MB17] Martin Beer, H. K. F. H. J. L. T. B. S. K. Martin Jahn J. Martin Jahn: Masterplan 100 Prozent Klimaschutz für die Landeshauptstadt Kiel. (2017)

[MBB11] Machowski, J.; Bialek, J.; Bumby, J.: *Power system dynamics: stability and control*. John Wiley & Sons, 2011

[MC11] Meng, F.; Chowdhury, B. H.: Distribution LMP-based economic operation for future smart grid. In: *2011 IEEE Power and Energy Conference at Illinois* IEEE, 2011, S. 1–5

[MCC05] Milano, F.; Cañizares, C. A.; Conejo, A. J.: Sensitivity-based security-constrained OPF market clearing model. In: *IEEE Transactions on power systems* 20 (2005), Nr. 4, S. 2051–2060

[MCC+13] Miret, J.; Camacho, A.; Castilla, M.; Vicuña, L. G.; Matas, J.: Control scheme with voltage support capability for distributed generation inverters under voltage sags. In: *IEEE Transactions on Power Electronics* 28 (2013), Nr. 11, S. 5252–5262

[MCS20] Machado, J. E.; Cucuzzella, M.; Scherpen, J.: Modeling and Passivity Properties of District Heating Systems. In: *arXiv preprint arXiv:2011.05419* (2020)

[MD17] Mattias Deutsch, F. S. D. S. B. A. S. P. S. D. D. S. Norman Gerhardt G. Norman Gerhardt: Wärmewende 2030: Schlüsseltechnologien zur Erreichung der mittel- und langfristigen Klimaschutzziele im Gebäudesektor / Agora Energiewende, Fraunhofer IBP, Fraunhofer IWES. 2017. – Forschungsbericht

[MDS+17] Molzahn, D. K.; Dörfler, F.; Sandberg, H.; Low, S. H.; Chakrabarti, S.; Baldick, R.; Lavaei, J.: A survey of distributed optimization and control algorithms for electric power systems. In: *IEEE Transactions on Smart Grid* 8 (2017), Nr. 6, S. 2941–2962

[MKP20] Mitridati, L.; Kazempour, J.; Pinson, P.: Heat and electricity market coordination: A scalable complementarity approach. In: *European Journal of Operational Research* 283 (2020), Nr. 3, S. 1107–1123

[MMW+20] Meng, X.; Mu, Y.; Wang, M.; Jia, H.; Huo, X.; Li, S.: Exploiting Flexibility of Heating Network for Integrated Energy System Considering Temperature Gradient and Fatigue Life. In: *2020 IEEE Sustainable Power and Energy Conference (iSPEC)* IEEE, 2020, S. 1204–1209

[MSC03] Ma, X.; Sun, D. I.; Cheung, K. W.: Evolution toward standardized market design. In: *IEEE Transactions on Power Systems* 18 (2003), Nr. 2, S. 460–469

[MVC17] Mhanna, S.; Verbic, G.; Chapman, A.: A Component-Based Dual Decomposition Method for the OPF Problem. (2017)

[NAB14] Nappu, M. B.; Arief, A.; Bansal, R. C.: Transmission management for congested power system: A review of concepts, technical challenges and development of a new methodology. In: *Renewable and Sustainable Energy Reviews* 38 (2014), S. 572–580

[NMZ+16] Nielsen, M. G.; Morales, J. M.; Zugno, M.; Pedersen, T. E.; Madsen, H.: Economic valuation of heat pumps and electric boilers in the Danish energy system. In: *Applied Energy* 167 (2016), S. 189–200

[Nou95] Noussair, C.: Equilibria in a multi-object uniform price sealed bid auction with multi-unit demands. In: *Economic Theory* 5 (1995), Nr. 2, S. 337–351

[NPC03] Nogales, F. J.; Prieto, F. J.; Conejo, A. J.: A decomposition methodology applied to the multi-area optimal power flow problem. In: *Annals of operations research* 120 (2003), Nr. 1-4, S. 99–116

[NTJK18] Nussbaumer, T.; Thalmann, S.; Jenni, A.; Ködel, J.: Planungshandbuch Fernwärme / EnergieSchweiz, Bundesamt für Energie BFE. 2018 (1.2). – Forschungsbericht

[NTJK20] Nussbaumer, T.; Thalmann, S.; Jenni, A.; Ködel, J.: Handbook on Planning of District Heating Networks / EnergieSchweiz, Bundesamt für Energie BFE. 2020 (1.2). – Forschungsbericht

[NW06] Nocedal, J.; Wright, S. J.: *Numerical optimization.* 2. ed. New York, NY: Springer, 2006

[OCS04] Overbye, T. J.; Cheng, X.; Sun, Y.: A comparison of the AC and DC power flow models for LMP calculations. In: *37th Annual Hawaii International Conference on System Sciences, 2004. Proceedings of the* IEEE, 2004, S. 9–pp

[Opp15] Oppelt, T.: *Modell zur Auslegung und Betriebsoptimierung von Nah-und Fernkältenetzen*, Diss., 2015

[Ott03] Ott, A. L.: Experience with PJM market operation, system design, and implementation. In: *IEEE Transactions on Power Systems* 18 (2003), Nr. 2, S. 528–534

[OUGP16] Oppelt, T.; Urbaneck, T.; Gross, U.; Platzer, B.: Dynamic thermo-hydraulic model of district cooling networks. In: *Applied Thermal Engineering* 102 (2016), S. 336–345

[PD11] Palensky, P.; Dietrich, D.: Demand side management: Demand response, intelligent energy systems, and smart loads. In: *IEEE transactions on industrial informatics* 7 (2011), Nr. 3, S. 381–388

[PKK15] Pillay, A.; Karthikeyan, S. P.; Kothari, D.: Congestion management in power systems–A review. In: *International Journal of Electrical Power & Energy Systems* 70 (2015), S. 83–90

[PSH20] Pelda, J.; Stelter, F.; Holler, S.: Potential of integrating industrial waste heat and solar thermal energy into district heating networks in Germany. In: *Energy* 203 (2020), S. 117812

[REN22] REN2022: *Renewables 2022 Global Status Report.* Paris: REN21 Secretariat, 2022 (ISBN 978-3-948393-04-5)

[RMIMD13] Rabiee, A.; Mohammadi-Ivatloo, B.; Moradi-Dalvand, M.: Fast dynamic economic power dispatch problems solution via optimality condition decomposition. In: *IEEE Transactions on Power Systems* 29 (2013), Nr. 2, S. 982–983

[Sau19] Sauter, P. S.: *Modellierung und zentrale prädiktive Regelung von multimodalen Energieverteilnetzen*, Karlsruher Institut für Technologie (KIT), Diss., 2019

[SBLS04] Son, Y. S.; Baldick, R.; Lee, K.-H.; Siddiqi, S.: Short-term electricity market auction game analysis: uniform and pay-as-bid pricing. In: *IEEE Transactions on Power Systems* 19 (2004), Nr. 4, S. 1990–1998

[SDPT16] Scholten, T.; De Persis, C.; Tesi, P.: Modeling and control of heat networks with storage: the single-producer multiple-consumer case. In: *IEEE Transactions on Control Systems Technology* 25 (2016), Nr. 2, S. 414–428

[SDWS17] Shao, C.; Ding, Y.; Wang, J.; Song, Y.: Modeling and integration of flexible demand in heat and electricity integrated energy system. In: *IEEE Transactions on Sustainable Energy* 9 (2017), Nr. 1, S. 361–370

[SGS+17] Sheng, T.; Guo, Q.; Sun, H.; Pan, Z.; Zhang, J.: Two-stage state estimation approach for combined heat and electric networks considering the dynamic property of pipelines. In: *Energy Procedia* 142 (2017), S. 3014–3019

[SMA22] *SMARD Strommarktdaten.* www.smard.de. Version: 2022

[SRG08a] Sensfuss, F.; Ragwitz, M.; Genoese, M.: The merit-order effect: A detailed analysis of the price effect of renewable electricity generation on spot market prices in Germany. In: *Energy Policy* 36 (2008), Nr. 8, S. 3086–3094

[SS12] Stephan Schweikardt, D. F. E. D. K. W. Michael Didycz D. Michael Didycz: Sektoruntersuchung Fernwärme / Bundeskartellamt. 2012. – Forschungsbericht

[SS18] Sarbu, I.; Sebarchievici, C.: A comprehensive review of thermal energy storage. In: *Sustainability* 10 (2018), Nr. 1, S. 191

[Sta] STADTWERKE, K.: *Zahlen zum Strom.* Online. https://www.stadtwerke-kiel.de/. – Accessed: 08.06.2022

[Str08] STRBAC, G.: Demand side management: Benefits and challenges. In: *Energy policy* 36 (2008), Nr. 12, S. 4419–4426

[STRW18] STAUDT, P.; TRÄRIS, Y.; RAUSCH, B.; WEINHARDT, C.: Predicting redispatch in the German electricity market using Information Systems based on Machine Learning. (2018)

[SWW+19] SUN, G.; WANG, W.; WU, Y.; HU, W.; YANG, Z.; WEI, Z.; ZANG, H.; CHEN, S.: A nonlinear analytical algorithm for predicting the probabilistic mass flow of a radial district heating network. In: *Energies* 12 (2019), Nr. 7, S. 1215

[SZP+09] STEVANOVIC, V. D.; ZIVKOVIC, B.; PRICA, S.; MASLOVARIC, B.; KARAMARKOVIC, V.; TRKULJA, V.: Prediction of thermal transients in district heating systems. In: *Energy Conversion and Management* 50 (2009), Nr. 9, S. 2167–2173

[Trö99] TRÖSTER, S.: *Zur Betriebsoptimierung in Kraft-Wärme-Kopplungssystemen unter Berücksichtigung der Speicherfähigkeit des Fernwärmenetzes.* Fraunhofer-IRB-Verlag, 1999

[TWW+20] TAN, J.; WU, Q.; WEI, W.; LIU, F.; LI, C.; ZHOU, B.: Decentralized robust energy and reserve Co-optimization for multiple integrated electricity and heating systems. In: *Energy* 205 (2020), S. 118040

[VDI13] *VDI-Wärmeatlas.* 11., bearb. und erw. Aufl. Berlin, Heidelberg: Springer Vieweg, 2013 (VDI-BuchSpringerLinkSpringer eBook Collection)

[Vic61] VICKREY, W.: Counterspeculation, Auctions, and Competitive Sealed Tenders. In: *The Journal of Finance* 16 (1961), Nr. 1, S. 8–37

[VSG+17] VERRILLI, F.; SRINIVASAN, S.; GAMBINO, G.; CANELLI, M.; HIMANKA, M.; VECCHIO, C. D.; SASSO, M.; GLIELMO, L.: Model Predictive Control-Based Optimal Operations of District Heating System With Thermal Energy Storage and Flexible Loads. In: *IEEE Transactions on Automation Science and Engineering* 14 (2017), April, Nr. 2, S. 547–557

[VSLD13] VESTERLUND, M.; SANDBERG, J.; LINDBLOM, B.; DAHL, J.: Evaluation of losses in district heating system, a case study. In: *International Conference on Efficiency, Cost, Optimization, Simulation and Environmental Impact of Energy Systems: 16/07/2013-19/07/2013*, 2013

[Wäc09] WÄCHTER, A.: Short tutorial: Getting started with ipopt in 90 minutes. In: *Dagstuhl Seminar Proceedings* Schloss Dagstuhl-Leibniz-Zentrum fuer Informatik, 2009

[WCHK18] WANG, Y.; CHEN, Q.; HONG, T.; KANG, C.: Review of smart meter data analytics: Applications, methodologies, and challenges. In: *IEEE Transactions on Smart Grid* 10 (2018), Nr. 3, S. 3125–3148

[WGJ+17] Wu, C.; Gu, W.; Jiang, P.; Li, Z.; Cai, H.; Li, B.: Combined economic dispatch considering the time-delay of district heating network and multi-regional indoor temperature control. In: *IEEE Transactions on Sustainable Energy* 9 (2017), Nr. 1, S. 118–127

[WJH+19] Wang, D.; Jia, H.; Hou, K.; Du, W.; Chen, N.; Wang, X.; Fan, M. u. a.: Integrated demand response in district electricity-heating network considering double auction retail energy market based on demand-side energy stations. In: *Applied Energy* 248 (2019), S. 656–678

[WYA+15] Wang, H.; Yin, W.; Abdollahi, E.; Lahdelma, R.; Jiao, W.: Modelling and optimization of CHP based district heating system with renewable energy production and energy storage. In: *Applied Energy* 159 (2015), S. 401–421

[WYJ+16] Wu, J.; Yan, J.; Jia, H.; Hatziargyriou, N.; Djilali, N.; Sun, H.: Integrated energy systems. In: *Applied Energy* 167 (2016), S. 155–157

[XWL14] Xu, J.; Wang, R. Z.; Li, Y.: A review of available technologies for seasonal thermal energy storage. In: *Solar energy* 103 (2014), S. 610–638

[XWZ+19] Xu, D.; Wu, Q.; Zhou, B.; Li, C.; Bai, L.; Huang, S.: Distributed multi-energy operation of coupled electricity, heating, and natural gas networks. In: *IEEE Transactions on Sustainable Energy* 11 (2019), Nr. 4, S. 2457–2469

[XZL+20] Xu, D.; Zhou, B.; Liu, N.; Wu, Q.; Voropai, N.; Li, C.; Barakhtenko, E.: Peer-to-Peer Multienergy and Communication Resource Trading for Interconnected Microgrids Microgrids. In: *IEEE Transactions on Industrial Informatics* 17 (2020), Nr. 4, S. 2522–2533

[YCC+19] Yu, Y.; Chen, H.; Chen, L.; Chen, C.; Wu, J.: Optimal operation of the combined heat and power system equipped with power-to-heat devices for the improvement of wind energy utilization. In: *Energy Science & Engineering* (2019)

[YHA+20] Yang, Z.; Hu, J.; Ai, X.; Wu, J.; Yang, G.: Transactive energy supported economic operation for multi-energy complementary microgrids. In: *IEEE Transactions on Smart Grid* 12 (2020), Nr. 1, S. 4–17

[YMS14] Yazdanian, M.; Mehrizi-Sani, A.: Distributed control techniques in microgrids. In: *IEEE Transactions on Smart Grid* 5 (2014), Nr. 6, S. 2901–2909

[YP21] Yu, M. G.; Pavlak, G. S.: Assessing the performance of uncertainty-aware transactive controls for building thermal energy storage systems. In: *Applied Energy* 282 (2021), S. 116103

[YS05a] Yao, R.; Steemers, K.: A method of formulating energy load profile for domestic buildings in the UK. In: *Energy and Buildings* 37 (2005), Nr. 6, S. 663 – 671

[ZBC04] Zareipour, H.; Bhattacharya, K.; Cañizares, C.: Distributed generation: current status and challenges, 2004

[ZGX13] Zhang, J.; Ge, B.; Xu, H.: An equivalent marginal cost-pricing model for the district heating market. In: *Energy Policy* 63 (2013), S. 1224–1232

[ZH21] Zheng, W.; Hill, D. J.: Incentive-based coordination mechanism for distributed operation of integrated electricity and heat systems. In: *Applied Energy* 285 (2021), S. 116373

[Zim10] Zimmerman, R. D.: Uniform price auctions and optimal power flow. In: *Matpower Technical Note, Feb* 1 (2010)

[ZLG13] Zhang, L.; Li, H.; Gudmundsson, O.: Comparison of district heating systems used in China and Denmark. In: *Euroheat and Power (English Edition)* 10 (2013), Nr. 4, S. 12–19

[ZLGZ17] Zhang, H.; Li, Y.; Gao, D. W.; Zhou, J.: Distributed optimal energy management for energy internet. In: *IEEE Transactions on Industrial Informatics* 13 (2017), Nr. 6, S. 3081–3097

[ZLZ+19] Zhou, H.; Li, Z.; Zheng, J.; Wu, Q.; Zhang, H.: Robust Scheduling of Integrated Electricity and Heating System Hedging Heating Network Uncertainties. In: *IEEE Transactions on Smart Grid* (2019)

[ZMS11] Zimmerman, R.; Murillo-Sanchez, C. E.: MATPOWER: Steady-State Operations, Planning, and Analysis Tools for Power Systems Research and Education. In: *Power Systems, IEEE Transactions on* 26 (2011), Feb, Nr. 1, S. 12–19

[ZWW+21] Zhang, M.; Wu, Q.; Wen, J.; Lin, Z.; Fang, F.; Chen, Q.: Optimal operation of integrated electricity and heat system: A review of modeling and solution methods. In: *Renewable and Sustainable Energy Reviews* 135 (2021), S. 110098

[ZXL+17] Zhong, W.; Xie, K.; Liu, Y.; Yang, C.; Xie, S.: Auction mechanisms for energy trading in multi-energy systems. In: *IEEE Transactions on industrial informatics* 14 (2017), Nr. 4, S. 1511–1521

[ZZZW18a] Zheng, J.; Zhou, Z.; Zhao, J.; Wang, J.: Effects of the operation regulation modes of district heating system on an integrated heat and power dispatch system for wind power integration. In: *Applied energy* 230 (2018), S. 1126–1139

[ZZZW18b] Zheng, J.; Zhou, Z.; Zhao, J.; Wang, J.: Integrated heat and power dispatch truly utilizing thermal inertia of district heating network for wind power integration. In: *Applied Energy* 211 (2018), S. 865 – 874

Own Publications and Conference Proceedings

[MEKH18] Maurer, J.; Elsner, C.; Krebs, S.; Hohmann, S.: Combined Optimization of District Heating and Electric Power Networks. In: *Energy Procedia* 149 (2018), S. 509–518

[MGR+21] Maurer, J.; Golla, A.; Richter, B.; Hohmann, S.; Weinhardt, C.: Hybrid Pricing Based Operation of Coupled Electric Power and District Heating Networks. In: *Sustainable Energy, Grids and Networks* 28 (2021), S. 100532

[MIJSH22] Maurer, J.; Illerhaus, J.; Jané Soneira, P.; Hohmann, S.: Distributed Optimization of District Heating Networks using Optimality Condition Decomposition. In: *Energies, Special Issue on Environmental Assessment and Optimization of Energy Systems and Technologies* 15 (2022), Nr. 18, S. 1–21

[MRMH21] Maurer, J.; Ratzel, O.; Malan, A. J.; Hohmann, S.: Comparison of Discrete Dynamic Pipeline Models for Operational Optimization of District Heating Networks. In: *Energy Reports* 7 (2021), S. 244–253

[MSKH16] Maurer, J.; Sauter, P.; Kluwe, M.; Hohmann, S.: Optimal energy management of low level multi-carrier distribution grids. In: *2016 IEEE International Conference on Power System Technology (POWERCON)*, IEEE PES, 2016, S. 1–6. – (Best Poster Award)

[MTK+22] Maurer, J.; Tschuch, N.; Krebs, S.; Bhattacharya, K.; Cañizares, C.; Hohmann, S.: Toward Transactive Control of Coupled Electric Power and District Heating Networks. In: *Applied Energy* (2022)

[SHM+19] Sauter, P. S.; Herkner, T.; Maurer, J.; Krebs, S.; Hohmann, S.: On Robustness Regarding Modeling Errors in Predictive Energy Management Systems for Multi-Carrier Energy Grids. In: *International ETG-Congress 2019; ETG Symposium, Esslingen, Germany, 8-9 May 2019*, VDE Verlag, 2019, S. 266–271

[ZSL+17] Zimmerlin, M.; Suriyah, M.; Leibfried, T.; Sauter, P.; Maurer, J.; Pfeifer, M.; Kluwe, M.; Hohmann, S.; Köppel, W.; Degünther, C.; Müller. C.; Ruf, J.; Prabhakaran, P.; Missal, P.; Walter, J.; Herold, L.; Eichinger. T.: Das Verbundprojekt RegEnKibo: Regionalisierung der Energieversorgung auf Verteilnetzebene unter Anwendung eines multimodalen, zellularen Ansatzes. In: *ETG Congress, Bonn, 28.-29.11.2017*, VDE Verlag, 2017. – 37.06.01; LK 01

Supervised Theses

[Els18] Elsner, C.: *Optimale Betriebsführung gekoppelter Strom- und Wärmenetze*, Institute of Control Systems (IRS), Karlsruhe Institute of Technology (KIT), Master's Thesis, 2018

[Fen19] Feng, Y.: *Erstellung eines Simulationsmodells zur Weiterentwicklung einer bestehenden Regelung für das doppelt-inverse Pendel*, Institute of Control Systems (IRS), Karlsruhe Institute of Technology (KIT), Bachelor's Thesis, 2019

[Gol18] Golla, A.: *Market and Control Based Coordination of Coupled Electric Power and District Heating Networks*, Institute of Information Systems and Marketing (IISM), Institute of Control Systems (IRS), Karlsruhe Institute of Technology (KIT), Master's Thesis, 2018

[Ide20] Idelhauser, S.: *Entwurf und Implementierung eines Dekompositionsansatzes von Optimierungsproblemen zur marktbasierten Betriebsführung von gekoppelten Strom- und Wärmenetzen*, Institute of Control Systems (IRS), Karlsruhe Institute of Technology (KIT), Master's Thesis, 2020

[Ill20] Illerhaus, J.: *Entwurf eines nichtlinearen Optimierungsproblems zur marktbasierten Betriebsführung von Wärmenetzen der 4. Generation*, Institute of Control Systems (IRS), Karlsruhe Institute of Technology (KIT), Bachelor's Thesis, 2020

[Rat19] Ratzel, O.: *Vergleich thermischer Modellierungsansätze von Wärmenetzen*, Institute of Control Systems (IRS), Karlsruhe Institute of Technology (KIT), Bachelor's Thesis, 2019

[Tsc19] Tschuch, N.: *Implementierung eines marktbasierten modellprädiktiven Reglers für ein gekoppeltes Strom- und Wärmenetz*, Institute of Control Systems (IRS), Karlsruhe Institute of Technology (KIT), Bachelor's Thesis, 2019

[Vie19] Vieth, J.: *Entwurf eines modellprädiktiven Reglers für das sektorengekoppelte Verteilnetz der Stadtwerke Kiel*, Institute of Control Systems (IRS), Karlsruhe Institute of Technology (KIT), Bachelor's Thesis, 2019

[Web17] Weber, D.: *Implementierung und Vergleich von Regelungsverfahren für das doppelt-inverse Pendel*, Institute of Control Systems (IRS), Karlsruhe Institute of Technology (KIT), Master's Thesis, 2017

[Wei18] Weiler, J.: *Implementierung und Vergleich von Ansätzen zur verteilten Berechnung des optimalen Leistungsflusses*, Institute of Control Systems (IRS), Karlsruhe Institute of Technology (KIT), Master's Thesis, 2018

Karlsruher Beiträge zur Regelungs- und Steuerungstechnik (ISSN 2511-6312)
Institut für Regelungs- und Steuerungssysteme

Band 01 Diehm, Gunter
Identifikation des menschlichen Bewegungsverhaltens auf der Basis von Primitiven.
ISBN 978-3-7315-0608-9

Band 02 Flad, Michael
Kooperative Regelungskonzepte auf Basis der Spieltheorie und deren Anwendung auf Fahrerassistenzsysteme.
ISBN 978-3-7315-0610-2

Band 03 Eckert, Marius
Modellbasierte Identifikation fraktionaler Systeme und ihre Anwendung auf die Lithium-Ionen-Zelle.
ISBN 978-3-7315-0690-4

Band 04 Krebs, Stefan
Intervallbeobachter für lineare parametervariante Systeme und deren Anwendung auf die Asynchronmaschine.
ISBN 978-3-7315-0857-1

Band 05 Kaspar, Stephan
Fahrdynamikuntersuchungen eines Elektrofahrzeugs mit Einzelrad-Hinterradantrieb.
ISBN 978-3-7315-0916-5

Band 06 Sauter, Patrick S.
Modellierung und zentrale prädiktive Regelung von multimodalen Energieverteilnetzen.
ISBN 978-3-7315-0963-9

Band 07 Kupper, Martin
Verteilte Zustandsschätzung fraktionaler Systeme und ihre Anwendung auf Lithium-Ionen-Batteriesysteme.
ISBN 978-3-7315-0971-4

Band 08 Merkert, Lennart
Optimal Scheduling of Combined Heat and Power Generation Considering Heating Grid Dynamics.
ISBN 978-3-7315-1056-7

Die Bände sind unter www.ksp.kit.edu als PDF frei verfügbar oder als Druckausgabe bestellbar.

Band 09 Ludwig, Julian
Automatisierte kooperative Transition einer Regelungsaufgabe zwischen Mensch und Maschine am Beispiel des hochautomatisierten Fahrens.
ISBN 978-3-7315-1069-7

Band 10 Inga Charaja, Juan Jairo
Inverse Dynamic Game Methods for Identification of Cooperative System Behavior.
ISBN 978-3-7315-1080-2

Band 11 Schnurr, Christoph Xaver
Ein Verfahren zur lexikographischen modellprädiktiven Regelung mit der Anwendung auf eine permanenterregte Synchronmaschine.
ISBN 978-3-7315-1095-6

Band 12 Schwab, Stefan
Guaranteed Verification of Dynamic Systems.
ISBN 978-3-7315-0965-3

Band 13 Pfeifer, Martin
Automated Model Generation and Observer Design for Interconnected Systems: A Port-Hamiltonian Approach.
ISBN 978-3-7315-1135-9

Band 14 König, Alexander
Absicherung hochautomatisierten Fahrens durch passiven virtuellen Dauerlauftest.
ISBN 978-3-7315-1141-0

Band 15 Stark, Oliver
Parameter- und Ordnungsidentifikation von fraktionalen Systemen mit einer Anwendung auf eine Lithium-Ionen-Batteriezelle.
ISBN 978-3-7315-1187-8

Band 16 Köpf, Florian
Adaptive Dynamic Programming: Solltrajektorienfolgeregelung und Konvergenzbedingungen.
ISBN 978-3-7315-1193-9

Band 17 Kölsch, Lukas
Dynamic Incentives for Optimal Control of Competitive Power Systems.
ISBN 978-3-7315-1209-7

Band 18 Schwartz, Manuel
Topologie-Optimierung eines radselektiv angesteuerten Fahrzeugs basierend auf einer optimalen Fahrzeugführungsregelung.
ISBN 978-3-7315-1222-6

Band 19 Rothfuß, Simon
Human-Machine Cooperative Decision Making.
ISBN 978-3-7315-1223-3

Band 20 Gellrich, Thomas Christoph
Dynamical Modeling and Control of Multiphase Heat Transport Systems Based on Loop Heat Pipes.
ISBN 978-3-7315-1231-8

Band 21 Maurer, Jona
Transactive Control of Coupled Electric Power and District Heating Networks.
ISBN 978-3-7315-1276-9